1980

# Conceiving the Self

# CONCEIVING
# THE SELF

Morris Rosenberg

*Basic Books, Inc., Publishers*

NEW YORK

Library of Congress Cataloging in Publication Data

Rosenberg, Morris.
  Conceiving the self.

  Bibliography:  p. 301
    1.  Self-perception.  2.  Self-respect.  3.  Self-
perception in children.  4.  Defense mechanisms (Psy-
chology)  I.  Title.  [DNLM:  1.  Self-concept.
BF697 R813c]
BF697. R657          155.2          78–19813
ISBN: 0–465–01352–X

TO THE MEMORY OF

EDWARD A. SUCHMAN

a stimulating colleague, a warm friend

# CONTENTS

# Contents

## PART III

### Self-Concept Development

## PART IV

### Beyond Self-Esteem

# PREFACE

IF there is one thing in the world that concerns every one of us, it is the self-concept. In the words of William James (1890: 289):

The altogether unique kind of interest which the human mind feels in those parts of creation which it can call *me* or *mine* may be a moral riddle, but it is a fundamental psychological fact. No mind can take the same interest in his neighbor's *me* as in his own. The neighbor's *me* falls together with all the rest of things in one foreign mass against which his own *me* stands out in startling relief.

Each of us knows, as a matter of immediate experience, that this "me" is something he cares about and that its protection and enhancement are major human motives.

This book is concerned with the self-concept—its nature (what it is), its social determinants (how society shapes it), and its development (how it changes over time). Our position is that the self-concept is the totality of the individual's thoughts and feelings with reference to himself as an object. Not only is this structure experienced as the core of the individual's interests but it also has major significance for his thoughts, feelings, and behavior.

Both popular and scholarly interest in the self-concept is probably higher today than ever before. This idea enters with increasing frequency into popular parlance, appearing in the press, on soap operas, in informal discussions. Whether one sees it as an idea whose time has come or as a faddish term destined for speedy extinction depends on one's personal crystal ball. In any case, it plays an important role in certain current ideological arguments, enters as a significant element in major policy decisions, and is increasingly invoked as an explanation for a variety of social ills. Furthermore, a remarkable variety of professional disciplines concerned with human behavior currently turn to the self-concept as an explanatory factor. Not only psychology, sociology, and psychiatry, but also education, nutrition, law, social work, medicine, nursing, political science, communication, athletics, human development, and many other fields are involved in self-concept research. But precisely because of this broadening interest in both lay and professional circles, it becomes all the more essential to have recourse to reasonably solid facts and straight thinking. Where these have been available, they have resulted in the puncturing of a number of widely held myths.

Although William James offered to the world his dazzling insights into the self-concept in the last century, and although the studies on this topic now number in the thousands, we believe that the surface has scarcely been scratched, the work hardly begun. Large gaps currently exist in our understanding of what the self-concept is, how it develops, and how it is shaped and molded by social forces. The present work attempts to deal with certain important issues which have either suffered undue neglect or which might be enriched by a somewhat different treatment. More specifically, it seeks to shed light on the following five questions:

1. What *is* the self-concept, that is, what does it include and how can one establish its boundaries?

2. What are the general *principles* of self-concept formation, that is, the theoretical presuppositions on which the great bulk of the empirical work rests?

3. How does social life—interpersonal interaction, immediate social context, broader social structure—help to shape the individual's views of what he is and wishes to be?

4. How does the self-concept (especially after the early years) change and develop as children grow older, not only in terms of general feelings about the self but also in terms of fundamental modes of conceptualization?

5. What mechanisms does the individual use to protect and defend his self-concept against the onslaughts that incessantly assail it?

In dealing with these issues, we shall not rely on a specific empirical study (though this book rests heavily on empirical data) nor shall we undertake a broad conceptual synthesis (though we shall make attempts at theoretical integration); rather, our intent is to present whatever theoretical and empirical material is best designed to deal with these questions. Most of the empirical research to be presented, it should be noted, focuses on the self-concepts of children and adolescents. There is good reason for this emphasis. It is in the preadult years that the self-concept emerges, evolves, and crystallizes; this is the time of life when the self-concept is most malleable, and when social and developmental factors operate in the most interesting, and sometimes unexpected, ways. But it is important to bear in mind that conclusions drawn from studies of children cannot necessarily be generalized to adults. As we shall see in chapter 5, children's and adults' self-concepts are formed by the same *principles* but not the same *influences*.

This book is divided into four parts, each beginning with a brief introduction. Part 1 deals with the first two questions specified above: what is the self-concept, what are the principles that operate to shape it. These two topics involve broad theoretical rather than empirical treatments. Chapter 1 is an extended discussion of the nature of the self-concept, considered in all its complexity. It directs attention to the multiple regions, components, and dimensions of this intricate structure. Chapter 2 deals with the major motives— self-esteem and self-consistency—associated with the self-concept, and advances four principles of self-concept formation. These principles serve as vantage points for approaching many of the substantive topics throughout this work.

Part 2 attempts to deal with the third question, namely, how diverse social influences operate to shape the self-concepts of children and adolescents. In contrast to the theoretical focus of part 1, part 2 (chapters 3–7) is based on the analysis of empirical data collected by the investigator and his collaborators. Chapter 3 starts with a consideration of the familiar idea of "significant others" (usually assumed to be parents, brothers and sisters, friends, classmates), introduced by the psychiatrist Harry Stack Sullivan. Although considerable attention has been given to this concept in the literature, what is often overlooked is that not all significant others are equally significant. Certain consequences of this fact are presented in this chapter.

In chapter 4, we move to somewhat broader interpersonal contexts—the school and the neighborhood. Like the family, these are contexts which the child had no hand in choosing. The focus of this chapter is on the relationship of the individual's characteristics and those prevalent in his environment. For example, much has been written on the significance of race or religion or social class for the child's self-concept. But the meaning and impact of these qualities depends on the interpersonal context in which they are embedded. It represents a different set of experiences to be a black in a white school than in a black school, or to be poor in a rich school than in a poor school. The bearing of such contextual dissonance or consonance on the child's self-concept is the theme of this chapter.

Chapters 5–7 turn to a consideration of the still broader social influences of social class and minority group status. Chapter 5 considers the question of the impact of the stratification system on the self-concept. Since social class represents (among other things) differential prestige in the society, it seems obvious that it should be related to self-esteem. But what seems obvious is not necessarily true. The empirical data in this regard are, in fact, extremely puzzling and yield highly inconsistent conclusions. Though the theory points in one direction and the facts in another, both, we believe, are true. In chapter 5, we argue that, although the identical *principles* account for the self-concepts of children and adults, social class, as it actually enters the experience of children and adults, produces different self-concept effects. In developing this point, we draw upon two bodies of empirical data: one, a sample of adults in the Chicago Urbanized Area, the other, a sample of public school children (grades 3–12) in Baltimore City.

Chapters 6 and 7 deal with minority group status—especially race and religion—and the self-concept. Probably more has been written on this subject than on any other in the sociology of the self-concept, and all of the theoretical writing has had the powerful support of everything but the facts. The idea that minority group children and adolescents tend to have lower self-esteem is, in the view of many writers, virtually an article of faith. Such firm convictions notwithstanding, we believe it is mistaken. In chapters 6 and 7, we develop the position that a number of unexamined assumptions underly the conclusion that minority group members have lower self-esteem, but that these assumptions frequently lack the support of facts.

Part 3 turns to a consideration of the fourth question on the preceding list, namely, how the self-concept changes and develops as children grow older. Three chapters are devoted to this relatively uncharted area. These data demonstrate that the view of the self-concept as permanently fixed in the early years is wildly inaccurate. Changes of a fundamental sort go on at least through middle childhood, early adolescence, and later adolescence. Chapter 8 examines the tendency among children of different ages to conceptualize the self either as a social exterior—a visible, overt, public object—or as a psychological interior—an inner world of thought, feeling, and experience. Chapter 9 turns to an examination of changes in other aspects of the self-concept—self-esteem, stability of the self-concept, self-consciousness, and content of the self-concept—presenting evidence that adolescent disturbance is no myth, and that it appears in *early* adolescence. Finally, chapter 10 reveals radical changes in the "locus of self-knowledge," dealing with the child's belief that the ultimate truth about the self is to be found within the individual or whether it is vested in some external source.

Finally, part 4 deals with the last question on the list, namely, how people go about protecting and defending their self-concepts in a world that constantly threatens them. One can only experience admiration, even awe, at the astonishing ingenuity exercised by the human mind—even the meanest intellect—to protect itself against attack. Although the defenses are not always successful, the work of self-protection is unceasing. The concluding chapter (chapter 11) details certain of these mechanisms.

Four basic perspectives underlie our treatment of these diverse topics. The first is our view that, if substantial progress in self-concept research is to be made, it is essential to go beyond self-esteem. Both in public discussions and in scientific research, it is common practice to equate the self-*concept* with self-*esteem*, thus failing to appreciate fully the richness, complexity, and explanatory power of the self-concept. This is not to suggest that self-esteem is unimportant; nothing could be farther from the truth. In fact, the major portion of this work is concerned with social and developmental factors involved in self-*esteem* formation. Furthermore, if one were to choose a single aspect of the self-concept for investigation, self-esteem might well be the most important. But what may be of *greatest* importance is not of *exclusive* importance. Hence, while paying considerable attention to self-esteem, we have at many places in this book (especially chapters 4, 7, 8, and 9) attempted to go beyond self-esteem, to deal with other aspects of the self-concept, many of major importance to individual feelings and behavior.

The second thread running through this work has been the consistent effort to examine *variables* within the framework of *principles*. Although each independent variable exercises its influence in its particular way, certain common principles underlie the dynamic processes through which these variables operate. For example, many investigators are interested in how the individual's groups, statuses, and social categories help to form and mold the self-concept. Cutting across these variables, however, is a small number of general principles—reflected appraisals, social comparisons, self-attribution, and psychological centrality—which help explain the common processes through which such variables as race, sex, age, or social class exercise their influence. Most of the data analysis in this book, then, while reflecting a serious interest in how each socio-demographic variable exercises its particular influence, at the same time is concerned with how the particular effect can be understood by reference to the abstract principles. In this work, we are interested in both the variables and the principles.

Third, we have attempted to view social forces as social experiences, and to keep in mind that these experiences do not simply exist but are interpreted. Social factors, after all, do not impinge on the self-concept (or, for that matter, any aspect of personality or behavior) *directly*. Class or race or religion or sex do not affect the self-concept; what they do is to subject the individual to a characteristic set of social and interpersonal experiences that are interpreted by participants in the situation. Race has never affected anyone's self-concept; only the experience of prejudice, discrimination, segregation, poverty can have this effect. Sex does not form a self-concept; being treated as a boy or girl, man or woman, does. Furthermore, these experiences, to a considerable extent, have their impact via the particular interpretation accorded them. The bearing of a racial slur on the feeling of self-worth depends on whether one accepts it as evidence of one's inferiority or as evidence of the irrationality, misguidedness, and possible psychopathology of the bigot. It is thus essential to understand how social structure is converted into concrete experience and is processed within the individual's phenomenal field. For this reason, we have attempted to follow W. I. Thomas' advice to see the situation from the viewpoints of the participants. This is no simple matter, especially when, as in this work, we are primarily studying children and adolescents, but their responses will remain forever mysterious unless we can adopt the child's-eye point of view.

The final perspective is the view of the individual as an active coper rather than as a passive object. The assumption that the individual docilely accepts the imprint of the various social forces bearing on him is as widespread as it is misleading. The individual is indeed influenced by social forces, but not in the sense of being molded by them but in the sense of responding, of dealing, of coping with them. The individual is directed by his own motivational system—self-esteem and self-consistency are two of these motives—and he responds to the structurally determined experiences of his life in a fashion protective of his self-concept.

# ACKNOWLEDGMENTS

SOME of the questions cited above have been dealt with in part in my previously published work; that material is repeated in this book (though usually in a modified form). Most of parts 1, 3, and 4, however, are presented for the first time, and large portions of the previously published material in part 2 are presented within a different framework.

Leonard I. Pearlin is co-author of chapter 5 and Roberta G. Simmons and Florence R. Rosenberg are co-authors of chapter 9. I am grateful to them for permitting me to publish the products of our joint efforts here. Roberta G. Simmons was also the co-investigator of the Baltimore school pupil study, upon which the present work draws heavily, and much of what appears throughout this book would never have been possible without her valuable contribution.

Since this work draws upon studies extending over many years, there are many people to thank. My first and foremost debt is to Erma Jean Surman, whose work on the New York State and the Baltimore studies, as research assistant, secretary, and friend, was brilliant and unflagging. Second, I have been the beneficiary of the excellent research assistance of B. Claire McCullough, whose outstanding statistical expertise and social-psychological knowledge proved invaluable in bringing the present work to fruition. I am also grateful to Jane Deiter, who displayed remarkable competence in typing the various drafts of an often unwieldly manuscript.

This work has benefited greatly from the suggestions and criticisms of the following people who read the manuscript in draft: Edward Z. Dager, Howard B. Kaplan, Melvin L. Kohn, B. Claire McCullough, Leonard I. Pearlin, Florence R. Rosenberg, and Roberta G. Simmons. I am also grateful to Midge Decter, senior editor at Basic Books, who provided constant encouragement and valuable advice in the course of this undertaking.

Portions of this work were completed during my term as Guest Worker at the Laboratory of Socioenvironmental Studies, NIMH. In addition, much of the data analysis was supported by Grant MH27747

## Acknowledgments

of the National Institute of Mental Health, ADAMHA. This support is deeply appreciated.

Grateful acknowledgment is made to the following publishers who granted permission to adapt some of my previously published material for use in this book:

American Sociological Association: *Black and White Self-Esteem: The Urban School Child*, Rose Monograph Series, 1972 (with R. G. Simmons); "Disturbance in the self-image at adolescence." *American Sociological Review* 38, 1973: 553–568 (with R. G. Simmons and F. Rosenberg).

Hemisphere Publishing Corporation: "The dissonant context and the adolescent self-concept." In S. Dragastin and G. Elder (eds.), *Adolescence in the Life Cycle*, 1975: 97–116.

Princeton University Press: *Society and the Adolescent Self-Image*, 1965: Table 5, p. 26; Table 9, p. 50; Table 13, p. 57; Table 5, p. 197; pages 243–248.

Sage Publications: "Which significant others?" *American Behavioral Scientist* 16, 1973: 829–860.

University of Chicago Press: "The dissonant religious context and emotional disturbance." *American Journal of Sociology* 68, 1962: 1–10; "Social class and self-esteem among children and adults." *American Journal of Sociology* 84, 1978: 53–77 (with Leonard I. Pearlin).

William Alanson White Psychiatric Foundation: "Contextual Dissonance Effects: Nature and Causes." *Psychiatry: Journal for the Study of Interpersonal Processes* 40, 1977: 205–217.

John Wiley and Sons: "Psychological selectivity in self-esteem formation." In C. Sherif and M. Sherif (eds.), *Attitude, Ego-Involvement and Change*, 1967: 26–50.

JAI Press: "Group rejection and self-rejection." Research in Community and Mental Health 1, 1978.

MORRIS ROSENBERG
February 1978
College Park, Maryland

# PART I

# The Self-Concept:
# Nature and Principles

I T IS SOMEWHAT ASTONISHING to think that after decades of theory and research on the self-concept, investigators are as far as ever from agreeing on what it is or what it includes. This is not to say that ambitious attempts at definition and specification have not been made. Indeed, as far back as 1890, William James presented an exposition on the nature of the self so insightful and so exhaustive that it remains unsurpassed to this day. It is no exaggeration to say that almost all the topics worth studying in this area were either adumbrated or clearly spelled out in James' classic discussion. Yet much of the subsequent work did more to obfuscate than to clarify the problem. Observing the incredible diversity of ideas that had accreted to the term "self," Gordon Allport suggested that the term be scrapped and that the more neutral Latin "proprium" be employed. His effort to distinguish and to specify the diverse areas of the proprium represented a major achievement in helping to create order out of chaos. And, in recent years, Chad Gordon (1968, 1974, 1976) has succeeded in specifying with remarkable detail and concreteness a major area of the self-concept, namely, its content and components.

Yet, the self-concept, in its full complexity, is still not adequately understood. Chapter 1—by far the longest in the book—attempts to grapple with this problem by spelling out in detail the various aspects, elements, or dimensions of the self-concept. This discussion is set forth not as the last word on the subject but as the first. Whatever the success of this effort, we are persuaded that until a reasonably clear idea of what the self-concept includes is gained, progress in self-concept research will be seriously impeded. In time, hopefully, the conceptual structure proposed in chapter 1 will be superseded by more refined and sophisticated analyses that will improve our empirical results by sharpening our thinking.

Chapter 2 deals with self-concept motives, and advances four principles of self-concept formation. Although not always stated explicitly, the theoretical foundation of most research in this area, we believe, rests on one or more of these principles: reflected appraisals, social comparison processes, self-attribution, and psychological centrality. Although these principles do not pretend to represent a full-fledged theoretical system, they help to make sense of many otherwise puzzling findings. When, in chapters 3–10, we turn to the presentation of empirical data from the two major empirical studies that form the bulk of this work (one, a study of high school juniors and seniors, the other, a sample of children from grades 3–12), these four principles will consistently serve as springboards for approaching the data.

# 1

## The Nature of the Self-Concept

THE HISTORY of self-concept research has followed a curious course. Although William James had written a brilliant chapter on "The Consciousness of Self" in his *Principles of Psychology*, nearly 60 years elapsed before the first systematic empirical research appeared (Raimy, 1948). The rivulet appearing at that time rapidly swelled to a stream, and soon became such a torrent that in the succeeding two decades nearly 2,000 studies on the subject were conducted (Gergen, 1971).

Despite the volume of work and the lapse of time, however, it is no exaggeration to say that we are as far as ever from agreeing on what it is, let alone on how to measure it. In fact, in a scientific field generally undistinguished by the precision of its terminology, the "self" stands as a concept foremost in the ranks of confusion. The substitution of related terms such as the ego (Sherif and Cantril, 1947), the proprium (Allport, 1955), and identity (Erikson, 1956) has not dispersed the clouds, mist, and vapors.

The terms "self" or "ego" have been used to refer to the "inner nature" or "essential nature" of man (Fromm, 1941, 1947; Maslow, 1954; Moustakas, 1956); to the experience and content of self-aware-

5

ness (Chein, 1944); to the individual as known to the individual (Hilgard, 1949; Murphy, 1947; Raimy, 1948; Rogers, 1951; Wylie, 1961, 1974); to a constellation of attitudes having reference to "I," "me," or "mine" experiences (James, 1890; Sherif and Cantril, 1947); to individual identity and continuity of personal character (Erikson, 1956); to a set of mental processes operating in the interest of satisfying inner drives (Freud, 1933; Symonds, 1951); and, most simply, to the person. There are, furthermore, various other distinctions or emphases in the literature. Turner (1976) speaks of "institutional" and "impulsive" selves; Franks and Marolla (1976) of "inner" and "outer" selves; Edelson and Jones (1954) of the "conceptual self-system"; Waterbor (1972) and Tiryakian (1968) of the "existential self" or the "existential bases of the self"; Seeman (1966) of "authentic" and "inauthentic" selves; Wylie (1961, 1968, 1974) and Snygg and Combs (1949) of "phenomenal" and "nonphenomenal" selves; Allport (1955) of the "proprium"; Sullivan (1947) of the "self-system"; Hilgard (1949) of the "inferred self"; and many others of the "self-image." [1] Wylie's (1968: 729) assertion that "There is no consistency in usage among theorists" shines forth as a euphemistic miracle; the less charitable might characterize the terminological situation as a shambles.

Over the years, however, one fundamental distinction has come to be recognized—that between the self as subject or agent and the self as object of the person's own knowledge and evaluation (Symonds, 1951; Hall and Lindzey, 1957; Wylie, 1974). The most immediate sense in which we experience the self is as an active agent—an executor, a doer. But it has long been recognized that one of the distinctive characteristics of the individual is his ability to serve as both subject and object simultaneously. This point has been expressed most vividly by George Herbert Mead, who indicated that the term "self" is a reflexive. Thus, such statements as "I have brown hair" or "I have a very moral personality" express a curious duality. The individual is standing outside himself and looking at an object, describing it, evaluating it, responding to it; but the object he is perceiving, evaluating, or responding to is himself. With regard to every other object in the world, the subject and object are different; only with respect to this object are they the same.

It should be noted that this ability is distinctive to mankind, one not shared by the lower forms of animal life. Thus it has been said that while all of animal life has consciousness, man alone has self-consciousness. Man alone can stand outside himself, describe himself, judge himself, condemn himself; man alone can feel pride, shame, guilt. The lion may be king of the jungle, but he puts on no airs before his jungle inferiors, any more than the rabbit blushes at his timidity.

In this work, our focus is on the self as object, not as subject—in other words, on the self-*concept*, not on the self. But what is this self-concept? When we use the term "self-concept," we shall mean *the totality of the individual's thoughts and feelings having reference to himself as an object.*[2]

But definitions rarely suffice to communicate complex ideas. What we mean by the self-concept may be clarified by indicating what, in our view, the self-concept is *not*.

First, the self-concept is not Freud's ego. As Symonds (1951) clearly indicated, Freud's ego consists of a set of intellectual processes enabling the individual to deal with reality, primarily serving the interests of the id (the fundamental unconscious wants of the individual) or placating the superego (conscience). The ego includes such cognitive processes as thinking, perceiving, remembering, reasoning, attending—all modes of thought used by the individual to gain his ends. To be sure, as Murphy (1947) and Allport (1955) observe, the ego also works to protect and enhance the self-concept, but it does not *constitute* the self-concept.

The self-concept is also not the "real self" (Horney, 1950), the "self-actualized person" (Maslow, 1954; Moustakas, 1956), the "productive personality" (Fromm, 1947), the "impulsive self" (Turner, 1976), or the "I" (Mead, 1934).[3] We do not mean to suggest that the individual's innate inclinations, dispositions, or drives are unimportant. A person may have a disposition to love, to write, to compose music, or to excel athletically, even though the conditions of life make it impossible for him to do these things. But these inner drives and inclinations represent a reality unequivocally distinct from the self-concept. The self-concept is not the "real self" but, rather, the *picture* of the self.

It is also important to distinguish the self-concept from "ego-involvements" (Sherif and Cantril, 1947); these ideas, although overlapping, are not identical. When an individual experiences pride or shame in some performance, group, object, or person having direct reference to the self, then the self-concept is implicated. But the term ego-involvement has also been extended to refer to other matters in which the individual may feel involved. Thus, a person may forget himself when involved in a task or a game, become lost in music or literature, give his all to a cause, or become totally absorbed in an interpersonal relationship; in these cases he may *lose* awareness of himself as an object (Smith, 1950). Although certainly *ego*-involved, his *self-concept* is not involved.

In recent years, considerable interest has been generated by Erikson's concept of "ego-identity." Erikson (1959: 102) himself acknowledged that he used the term in different senses: sometimes as "a conscious sense of *individual identity;*" at other times as "an unconscious striving for *continuity of personal character;*" at others "as a criterion for the silent doings of *ego synthesis;*" and at still others "as a maintenance of an inner *solidarity* with a group's ideals and identity." Although some of these elements fall under the rubric of the self-concept, Erikson's interests largely diverge from the cognitive emphasis involved in our definition.

Nor is the self-concept the existential self. The existential self refers to the world of being, of existence, of immediate experience. It is concerned with what is, not what has been or will be—with being, not becoming. The self-concept, on the other hand, is an object of perception and reflection, including the emotional responses to that perception and reflection. It is a product of "self-objectification," requiring the individual to stand outside himself and to react to himself as a detached object of observation.

Finally, some writers apparently use the term "self" to reflect the total constellation of an individual's psychological characteristics—a concept usually subsumed under the term "personality"—or even to characterize the total person. However important this topic may be, the self-concept is only one part of the individual's total personality and a still smaller part of his total person.

In defining what the self-concept is not, it may appear that we have defined away everything of interest, achieving clarity at the expense of significance. Nothing could be farther from the truth. Our meaning of the term self-concept—the totality of the individual's thoughts and feelings with reference to himself as an object—covers a great deal indeed. It might be more appropriate to refer to this area of interest as "the realm of self-ideas," as the individual's *Selbstanschauung*—his general guiding self-views. It is a concept with breadth and depth, one with profound consequences and ramifications both for the individual and society.

But this definition is assuredly unspecific. In attempting to spell out just what does suitably fit under the rubric of the "self-concept," it is convenient to distinguish three broad regions: the extant self (how the individual sees himself); the desired self (how he would *like* to see himself); and the presenting self (how he shows himself to others).

## Extant Self-Concept

What does the individual see when he looks at himself? Any reasonably complete description of the extant self-concept must take account of at least four areas. It must consider, first, the parts (content of the self); second, the relationship among the parts (structure); third, the ways of describing both parts and whole (dimensions); and finally, the boundaries of the object (ego-extensions).

CONTENT OF THE SELF-CONCEPT

The parts, elements, or components of the self-concept consist primarily of social identity elements, dispositions, and physical characteristics. Whereas the social identity elements are usually expressed in the language of nouns, dispositions or traits are usually expressed in the language of adjectives, although descriptive phrases, including verbs, can be satisfactorily substituted. Let us consider first the elements most fundamental to the sociologist—the elements of social identity.

*Social Identity.* It is characteristic of the human mind to classify the parts of reality that enter its experience into categories, and this applies to people as well as to other objects. Scarcely has an infant entered the world before it is classified in terms of sex (male or female), age (infant), race (black, white, oriental), nationality (American, Italian), religion (Protestant, Catholic, Jew), family status (son, daughter, niece, nephew, grandchild), legal status (legitimate, illegitimate), name (including family affiliation), and so on. Still a social neuter, the infant is immediately tagged, labeled, and pigeonholed in terms of socially defined categories (some of the precoded categories are even on the birth certificate). In the course of life new bases of classification are added: the individual becomes a doctor, lawyer, machinist; a Democrat or Republican; a Mason, Shriner, or Elk.

It is these categories which constitute the individual's social identity—that is, the groups, statuses, or categories to which he is socially recognized as belonging. In one sense these elements are easily identified, for they are expressed in a language of *nouns*. Without laying claim to exhaustiveness, six categories of social identity merit attention: social statuses, membership groups; "labels"; derived statuses; types; and personal identity.

Social statuses are universal bases of social classification and self-definition. Sex, age, family status, stratification position, and occupation are criteria of classification in all societies.

As societies grow larger, they subdivide into groups based on voluntary association, similarity of belief or interest, sharing of culture or subculture, commonality of origin, physical or regional contiguity, or whatever. These are the individual's *membership groups*. These might involve a full-fledged cultural identity, such as a Frenchman, Italian, Greek (based on common language, shared norms and beliefs, common history and tradition, shared heroes, geographical closeness and common territory); groups based on common beliefs, such as religious groups (Catholic, Jewish, Baptist, Presbyterian), sociopolitical groups (Democrat, Republican, Gulliver's "big enders" and "little enders"), fellowship groups (Shriners, Masons, Elks), interest groups (union members, American Sociological Association members), and certain other socially defined categories, such as race (white, black, oriental). These membership groups are socially recognized bases of classification and constitute important elements of social identity.

A third aspect of social identity is based on the process of "social labeling." When a person behaves in a fashion contrary to social norms, his behavior is defined as deviant. Action as such is usually represented by a verb—*stealing* a bike, *drinking* too heavily, *taking* drugs, *engaging* in homosexual acts. It is only when society, especially through some certifying agency—a judge, a doctor, a political unit—locates such actors in defined and socially recognized categories that labeling occurs. The individual is no longer someone who drinks heavily but an *alcoholic*; who takes drugs, but a *drug addict*; who behaves peculiarly, but a *mental case*; who has shoplifted, but a *criminal*. When the language of verbs becomes a language of nouns, either through formal certification procedures or general social recognition, the labeling process occurs, and produces additional elements of social identity.

A fourth type of classification, usually rooted in the individual's history or biography, is essentially derived from other statuses, membership groups, or labels. Thus, a person may be classified as an ex-convict, ex-alcoholic, war veteran, emeritus professor, Harvard alumnus, ex-mental patient, divorcee, lame-duck Congressman. Whether the individual likes it or not, as far as society is concerned part of what he is is what he was. These categories affect his life in numerous ways, whether he is involved in concealing the status (ex-convict) or in gaining current recognition on grounds of past achievement (emeritus professor).

More vague and less general than the above forms of categorization are what might be called *social types*. These will usually be some syndrome of interests, attitudes, characteristics, or habits which are socially perceived as hanging together. In the broader society a person may be characterized as an intellectual, a playboy, a jet-setter, a culture bug, a Uriah Heep, or a Don Juan. Within narrower environments, such as the high school, the subculture may recognize various types—the jock, brain, big wheel, hood, freak, goof-off, greaser, loser, grind, eager beaver. Although recognition of such types tends to be confined within narrower subcultures (see Strong, 1942; Gordon, 1976), these types may nevertheless become salient aspects of the individual's social identity.

These types are extremely varied and defy simple, orderly classification. People may see themselves as natural leaders, world-weary sophisticates, charming hostesses, blasé men of the world, average Joes, the life of the party, artistic spirits, and so on. The place of these types in the self-concept is complex, for they may refer to how the individual sees himself, how he wishes to see himself, or how he seeks to appear to others.

The final component of social identity is, paradoxically, personal identity. The term "personal identity," of course, is often used to refer to the individual's deepest thoughts, feelings, and wishes, but that is not our present meaning. As here used, personal identity is also a matter of social classification but involves classifying the individual into a category with one case. It is best expressed by assigning the individual a unique label, usually a name. These socially assigned tags or labels are only as complex as needed for the purpose.

In a small group, Tom, Will, Jim may suffice; in a larger group (or one in which kinship is significant), Thomas Johnson (Thomas, son of John) is required for such identification. Today, even this refinement is insufficient, and the personal identity of the individual is increasingly expressed in a social security number or set of fingerprints.

It is thus apparent that the elements of social identity are numerous and are experienced as important parts of the self-concept. It is not simply that the individual *has* a sex, *belongs* to a race, or *believes* in the Democratic party; rather, in his view, he *is* a male, a black, a Democrat. An outstanding demonstration of this fact appears when people are asked to answer the general question, "Who am I?" in 20 words or phrases (Kuhn and McPartland, 1954). The first answers tend to be in terms of the elements of social identity—student, girl, husband, Baptist, Puerto Rican, and so on. These elements of social identity are part of the self-concept, in some cases constituting its hardest core. In a very real sense the individual *is* his social structure, for these socially recognized categories are a firm, deep, and real part of what he feels himself to be.

Certain distinctive characteristics of these social identity elements should be noted. First, it cannot be emphasized too strongly that these elements are *socially* defined; they are not "logical," "natural," or "scientific" categories. At first glance the division of people into categories may seem natural—they "cleave naturally at the joints," as Plato said—but this is not always the case. By what logic is a person who is 15/16 white classified as black in our society? [4] On what principle is a person without religious conviction classified as Catholic or Jewish? Similarly, although sex is a natural division, role assignment on the basis of it is by no means necessary. Nor are the identity elements necessarily *scientific* classifications. The scientist, of course, is free to classify people as bourgeoisie and proletariat, as endomorphic, mesomorphic, and ectomorphic, as status consistent and status inconsistent, but if these are not *socially* recognized bases of classification, then they do not constitute social identity elements.

Second, the broader society *responds* to people not so much in terms of what they are as in terms of these categories. People may look with respect on a professor (though he be a dunce), show disdain toward a black (though he be a genius), express fear and distrust of an ex-convict (though he be scrupulously honest), feel uneasiness toward a released mental patient; and so on. In other words society builds up a set of *social expectations* toward, or stereotypes of, people classified in terms of social identity elements; people base their behavior toward the individual on these "typifications" (Schutz, 1970; Berger and Luckmann, 1966).

These social identity elements, it should be stressed, are *membership* groups or categories, and may or may not represent reference groups. The individual may disidentify with the group, status, or social category—that is, he may firmly reject the group and wish he were not part of it—but he has no choice but to recognize himself as a member if he is so defined by society. A Jew, a black, or a member of the lower class may wish he belonged to another group or category, but he still perceives himself in terms of the social identity element.

The third point, of critical importance in the present context, is that many of these elements are socially ranked and evaluated; hence, the individual's sense of *personal worth* or value is to some extent contingent upon the prestige of the identity elements. The rank order of racial, religious, and ethnic groups has, with rare historical exceptions, been remarkably consistent over decades (Bogardus, 1928, 1959). Anglo-Saxon Protestants of British, Scottish, or Welsh descent have generally ranked at the top, followed by other Northern European groups, Southern Europeans, Jews, etc.; blacks have consistently ranked low or at the bottom. The rank order of occupations is equally consistent. Between 1947 and 1963, the rank order of occupational desirability for 90 selected occupations in a national sample remained virtually unchanged (Hodge, Siegel, and Rossi, 1964). Social classes, too, are by definition evaluated as better or worse, higher or lower, as are certain other groups.

Although one cannot assume a direct conversion of social identity evaluation into self-evaluation, one can assume that people *respond* in various ways to the social evaluation of their identity elements. So central is the evaluative component that labeling theorists have integrated it into the definition of the concept. To characterize someone as a convict, alcoholic, drug addict, radical, or mental patient (or even as an "ex" of these) is to stigmatize him, to brand him as unworthy, to provide him, in Goffman's (1963) terms, with a "spoiled identity."

Finally, it is worth noting that the social identity element is more than merely a convenient category or pigeonhole into which individuals are placed; rather, it frequently includes a social model serving as a standard for self-assessment. It is not simply that a woman sees herself as a doctor, a Catholic, or a mother but that she is concerned with whether she is a "good" doctor, Catholic, or mother, that is, with whether she matches some socially recognized role ideal. This ideal contains many elements such as traits, attitudes, behavior, values, norms, and codes of honor. An obvious example of such a role model is the image of the physician projected on the TV screen: dedicated, hard-working, kindly, intelligent, upstanding. The role model of the mother in American society is of someone who is kind, sympathetic, loving, helpful, self-sacrificing. These role models represent standards to which the status incumbent's role behavior is compared. To some extent, the individual's feeling of personal worth may hinge on the degree to which he lives up to the role model. A nurse may feel that she should always feel sympathetic toward the ill, experiencing strong guilt when this proves impossible, as it must. A mother may have deep feelings of self-hatred when she finds herself expressing, or even feeling, hostility toward her child. A doctor may believe that he should always help the suffering and feel great chagrin when he finds himself unable to do so. In all these cases, people may experience the same emotions—guilt, self-hatred, self-contempt—because they have fallen short of the culturally defined role ideal which they have internalized. Conversely, of course, the individual may feel great satisfaction at the adequate performance of his role and take pride in outstanding performance.

In addition to desired modes of *behavior*, people have conceptions of the *traits* appropriate to the role—the social worker as sympathetic, the sergeant as gruff, the artist as free-spirited—and they cultivate these traits, using them as standards for self-judgment. Although this model applies most clearly to the social roles connected with statuses, the same principle applies, to some extent, to many other social identity elements. In general, certain patterns of behavior and traits of character are associated with most social identity elements—man, woman, black, white, Mexican-American, child, adolescent, adult, doctor, lawyer, professor. The person perceives and evaluates himself in terms of the characteristics socially expected of a member of that category, and seeks to convert himself into the appropriate self by cultivating the approved characteristics.

Whether the identity element is positive or negative, then, the individual is aware of, and reacts to, the fact that a whole set of behavioral and personality characteristics are socially expected of someone carrying a certain tag. These characteristics serve as standards for self-evaluation, and importantly affect how the individual eventually comes to perceive, judge, and feel about himself.

*Dispositions.* Since consciousness precedes self-consciousness, the self-concept is something that emerges and develops gradually, primarily out of social experience. When the first glimmer of the self-idea enters the mind of the infant, it does so in the form of perceptions of concrete reality, not of abstract ideas. The self perceived by the baby may consist of his fingers, toes, belly, eyes, head, and so on. With increasing maturation and social learning, however, the child comes to view the objects of the world, including himself, in terms of abstract categories. He comes to see himself largely as a person with certain tendencies to respond, that is, with dispositions (Flavell, 1974). These may refer to attitudes (liberalism, conservatism), traits (bravery, generosity, morality), abilities (musical skill, athletic prowess, intelligence), values (belief in democracy, equality, success), "personality traits" (compulsiveness, extroversion), specific habits or acts (brushing teeth daily, working five days a week), likes or preferences (likes tennis, enjoys fine wine), or to "powers" or "tendencies" generally.

To a considerable extent, the individual views himself not as such but in terms of these abstract categories. He sees himself as liberal, intelligent, kindly, brave; good as a raconteur, knowledgeable about rare stamps, skilled at pole vaulting; as outgoing, friendly, anxious, optimistic; and so on. The content of the self-concept is largely made up of these abstract qualities. Although characteristically expressed as adjectives, these components of the self-concept do not hinge on the mastery of a particular vocabulary. The individual can think of himself as diligent or imitative or reliable without having mastered these particular terms.

Perhaps nothing more vividly demonstrates the complexity of the self-concept than the huge number of dispositions that may enter into it. Many years ago Allport and Odbert (1936) listed and classified all the adjectives they could find in an unabridged dictionary; the total was 17,958. Not all of these adjectives, to be sure, are applicable to human beings, but a surprising number are—all ways in which the individual can potentially characterize himself abstractly. If one adds to these the endless possible behavioral tendencies, talents or abilities, tastes and interests, attitudes and values, and inner thoughts and feelings, we can see how very numerous the individual's dispositions may be. The most sophisticated classification of self-components is Gordon's "Person-Conceptions (PC) System" which is built around a computer-readable dictionary of some 12,000 distinct person-descriptive meanings, grouped under 250 detailed categories (Gordon, 1976: 406), although finer differentiations raise the number to approximately 20,000 (Gordon, 1974).

Although dispositions and social identity elements (along with physical characteristics) constitute the bulk of the content of the self-concept, they differ in certain important ways. For one thing, whereas dispositions may be felt as more of what we *truly* are, the identity elements tend to be experienced as more of what we *surely* are. The individual, for example, may feel that the social identity elements represent solely his social exterior whereas the "real me" is expressed in his dispositions. The world, he feels, looks upon him as a lawyer, a black, an American, a Catholic, but the real self—what he deep down truly and really is—is sensitive, poetic, gentle, and philosophical. It is interesting, however, that the self-concept component of which he is most *certain* is the social identity element. Although there may be some question in his mind about how sensitive, intelligent, musically talented, or knowledgeable he is, there is no question about whether he is a lawyer, a black, a Catholic, or an American.

This is not to say that all social identity elements are firm (there are ambiguous statuses such as adolescent, oldster, and upper-middle class member), nor are all dispositions equivocal—far from it. In general, however, identity elements are more definite and unchangeable aspects of the self-concept than are dispositions.

In part, this difference is attributable to other characteristics distinguishing the individual's identity elements from his dispositions, namely, their *accuracy* and *verifiability*. The individual's perceptions of his socially recognized statuses, groups, or social categories are largely accurate because they are matters of social definition, classification, and recognition which are constantly verified in the social world; that is, people behave toward him as a black, a doctor, or an adolescent. Hence, with regard to most social identity elements, there is little room for argument or dispute. With reference to dispositions, however, people may not see themselves as they really are, nor do others necessarily see them as they see themselves; the certainty that comes from consensual validation is lacking. We shall return to this point in our discussion of the self-concept dimensions.

In addition to having certain conceptions of his dispositions and social identity elements, the individual has a picture of his various physical characteristics. Unfortunately, there is considerable disagreement in the literature on the meaning and significance of the "body-image." [5] According to Schilder (1968:107): "The image of the human body means the picture of our own body that we form in our mind, that is to say the way in which the body appears to ourselves." This definition corresponds to our meaning, although Schilder has in mind a concept considerably broader than physical characteristics, for he sees the individual's body-image as integrally connected with the body-images of other people. We have opted for the term "physical characteristics" rather than the better-known "body-image" in order to stress our interest in the physical self as a perceptual object. Such physical characteristics become especially interesting when they are substantially at variance with objective facts about the self (such as height, weight, or body build), or in conflict with most other people's perceptions of these physical qualities. The prominence of such physical characteristics in the individual's self-concept, we shall see in chapter 8, is strongly influenced by developmental factors.

STRUCTURE OF THE SELF-CONCEPT

In speaking of the content of the extant self-concept, we are referring to the major parts, components, or elements of the self-concept. But we cannot understand the extant self-concept without considering the *relationship* among the components. Some investigators implicitly treat the elements of the self-concept as items in a laundry list, as soldiers in a rank, neatly lined up in arbitrary order. To others, the individual's phenomenal field appears to consist of randomly scattered elements, wastebasket fragments scattered on the floor, flotsam and jetsam on the cognitive beach: such descriptive terms as generous, witty, nephew, Buddhist, football fan are viewed as strewn carelessly about the phenomenal field.

Such implicit assumptions do serious violence to the reality of the self-concept. In the words of Combs and Syngg (1959: 126): "The phenomenal self is not a mere conglomeration or addition of isolated concepts of self, but a patterned interrelationship or Gestalt of all these." Some elements of an attitude structure are central, others peripheral; some congeal into larger wholes (as in types), others are detached, standing in splendid isolation. It is not just the parts, but also the relationship among the parts, that constitutes the whole.

It must be acknowledged that little is currently known about this structure. Yet we believe that some purchase on the issue can be gained by recognizing three points: (1) that the self-concept components are of unequal centrality to the individual's concerns and are hierarchically organized in a system of self-values; (2) that the self-concept can be viewed at both the specific and global levels; and (3) that the self-concept may consist primarily of a social exterior or of a psychological interior.

*Psychological Centrality.* Self-values afford an apt illustration of the relevance of self-concept structure for the individual's global self-esteem. By self-values we refer to conceptions of the desirable which serve as standards or criteria for self-judgment. Ordinarily, we assume that if someone respects himself in certain particulars, then he respects himself in general. If the thinks he is smart, attractive, likeable, moral, interesting, and so on, then he thinks well of himself in general. Alternatively, if he does not feel he lives up to these characteristics, he thinks poorly of himself. This is clearly the assumption behind the Adjective Check List and kindred measures, and it also underlies the common assumption that minority group members, school failures, low socioeconomic status members, delinquents, or mental patients must have low self-esteem.

Yet it should be apparent that the significance of a particular component depends on its location in the self-concept structure—whether it is central or peripheral, cardinal or secondary, a major or minor part of the self. In other words, a person's global self-esteem is based not solely on an assessment of his constituent qualities but on an assessment of the qualities that *count*. This point was emphasized with characteristic felicity by William James (1890: 309):

I am often confronted by the necessity of standing by one of my empirical selves and relinquishing the rest. Not that I would not, if I could, be both handsome and fat and well-dressed, and a great athlete, and make a million a year, be a wit, a *bon-vivant*, and lady-killer, as well as a philosopher; a philanthropist, statesman, warrior, and African explorer, as well as a "tone-poet" and saint. But the thing is simply impossible. . . . So the seeker of his truest, strongest, deepest self must review this list carefully, and pick out the one on which to stake his salvation. . . . I, who for the time have staked my all on being a psychologist, am mortified if others know much more psychology than I. But I am contented to wallow in the grossest ignorance of Greek. My deficiencies there give me no sense of personal humiliation at all. Had I "pretensions" to be linguist, it would have been just the reverse.

The differential importance of self-concept components is thus critically significant for self-esteem. Some dispositions or social identity elements rank high in our hierarchy of values—stand at the center of our feelings of worth—whereas others are relegated to the periphery. One person stakes himself on his intelligence but cares little about his savoir faire; for another the reverse is the case. One takes great pride in his social class position, a second in his ethnic background, a third in his race, a fourth in his religious affiliation. A professor may consider himself both intelligent and well-mannered —both dispositions, in other words, are elements of his self-concept— but he may stake himself far more heavily on the former characteristic than on the latter.

Perhaps the most vivid expression of the unequal importance of identity elements in the individual's phenomenal field is expressed in the labeling theorist's notion of "role engulfment." When the labeling theorist uses this term, he essentially means that the deviant social identity achieves overwhelming prominence in the life of the individual. The fact of being a convict, mental patient, or alcoholic (or, indeed, of being an ex-convict, ex-mental patient, ex-alcoholic) becomes the central aspect of the individual's social and self-identity. The fact that he is professional, Protestant, handsome, a good father, well-mannered, interesting—all this is as nothing compared with his hatred of himself as an alcoholic, a homosexual, or an embezzler.

There is, however, no reason to restrict the concept of role engulfment to elements of deviant identity; the same may be true of any disposition or identity element which achieves pivotal importance in a person's mind. For some people being black may be all important; for others being a good Jew or Catholic may be of highest significance. Social types may also dominate the minds of people. A person may focus his pride system on his self-concept as an "intellectual" who has a typical set of interests, attitudes, values, and habits. Another may see himself as a "doer"—a person who cuts through red tape, brushes aside opposition, gets results. As we shall see in later analyses (especially chapter 7), failure to take account of this point has been responsible for a number of misleading inferences on the part of certain theorists.

*Part-Whole Relationships.* If one aspect of self-concept structure deals with the arrangement of the parts vis-à-vis one another, a different aspect concerns the relationship of the parts to the whole. Currently no consensus exists among self-concept researchers about whether it is preferable to focus attention on the parts—dispositions, identity elements, or physical characteristics—or on the whole— global self-attitudes. Indeed, it has been questioned whether there *is* any such thing as a global self-attitude, essentially on two grounds. First, it is obvious that there may be inconsistency among diverse self-concept components: the individual may think very highly of his intelligence, fairly highly of his morality, very poorly of his attractiveness; may feel pride in his social class, shame in his ethnic heritage; and so on. Given such variability, is there such a thing as an overall self-attitude? Second, the situational variability in the self-concept is undeniable; the person feels self-satisfaction at certain times and under certain conditions, self-dissatisfaction at others.

Although inconsistency or variability of self-attitudes may exist, we believe that the self-concept is not essentially different from attitudes toward any other object in the world. A person may have global positive or negative attitudes toward State University as a whole, and he may also have positive or negative attitudes toward the various parts of the university: he enjoys the lectures, considers the library facilities inadequate, revels in the company of his dormitory companions, finds the food pallid, and so on. There is surely nothing inconsistent in holding attitudes toward both the object as a whole as well as toward the component elements.

The question of whether it is preferable to focus attention on the global self-attitude or whether, on the contrary, the parts should be of nuclear interest is thus easily answered: in our view, both are legitimate and important areas of self-concept research. The chief point to stress, however, is that they are not identical or interchangeable: both exist within the individual's phenomenal field as separate and distinguishable entities, and each can and should be studied in its own right.

Failure to recognize this evident point has, we believe, led to a number of misleading inferences in the literature. For example, the great majority of studies offering evidence that the individual has negative feelings toward his skin color or even his race (Clark and Clark, 1947; Goodman, 1952; Brody, 1964; Kardiner and Ovesey, 1951; Proshansky and Newton, 1968) have concluded that the subject has lower global self-esteem. Various policies have been adopted in the past in the name of this conclusion. But it is logically and empirically unwarranted. Attitudes toward one's skin color is one thing, attitudes toward oneself as a whole is another. Both are important but they are neither equivalent nor identical.

If it be acknowledged that the parts and the whole are separate and distinct, then it follows that one cannot safely generalize from the specific to the global, or vice versa. For example, some educational programs designed to improve the academic performance of disprivileged children have attributed poor achievement to the child's belief that he is not smart enough to learn and have advocated measures calculated to improve the child's global self-esteem. But this logic assumes that the general is transferable to the specific, and that a change in the global self-attitude will produce a corresponding change in the specific self-attitude. This assumption is unjustified. The assessment of one's academic ability and the view of one's general self-worth are two separate attitudes whose relationship must be investigated, not assumed. For example, Bachman (1970: 242) shows that the correlation of the "self-concept of school ability" and global self-esteem is .33. Although the specific and general obviously overlap, they are also far from identical.

The question thus arises: what is the relationship between the whole and the parts? Since any totality is in some sense made up of the parts that constitute it, many researchers proceed by acquiring information about the individual's attitude toward his specific characteristics and adding up the responses in order to arrive at a global self-esteem score.[6] The critical drawback to this procedure is that it overlooks the extent to which the self-concept is a structure whose elements are arranged in a complex hierarchical order. Hence, simply to add up the parts in order to assess the whole is to ignore the fact that the global attitude is the product of an enormously *complex synthesis* of elements which goes on in the individual's phenomenal field. It is not simply the elements per se but their relationship, weighting, and combination that is responsible for the final outcome. The subject himself may be as ignorant as the investigator about how this complex synthesis of elements has been achieved, but he is in a unique position to recognize, as a matter of immediate experience, the final result. He alone can experience whether he has a generally favorable or unfavorable, positive or negative, pro or con feeling toward himself as a whole.

The danger of generalizing from the assessment of the part to the whole lies in the fact that the self-concept components on which the individual may elect to focus his sense of worth are remarkably diverse. Speaking of the neurotic, Horney notes: "There is simply nothing that may not be invested with pride. What is a shining asset to one person is a disgraceful liability to the next. One person is proud of being rude to people; another is ashamed of anything that could be construed as rudeness and is proud of his sensitiveness to others. One is proud of his ability to bluff his way through life; still another is ashamed of any trace of bluffing. Here is one who is proud of trusting people and there is one equally proud of distrusting them —and so forth and so on" (1950: 93).

What holds true for dispositions obtains equally for social identity elements and ego-extensions. People may be proud of being a third-generation American or a member of the country club, of having a fancy car or living in a good neighborhood, of giving successful parties or wearing the best clothes, and so on. What individuals and groups succeed in investing their pride in thus defies description. Parts and whole are both important, but they are neither transferable, interchangeable, nor equivalent, nor can information on the one form the basis for conclusions about the other.

A third aspect of self-concept structure deals with the conceptualization of the self in terms of a social exterior or a psychological interior. Some people show a propensity to view the self in terms of a social exterior—an overt, visible, palpable self—including physical characteristics, social identity elements, concrete behavior, and related components, whereas others are more inclined to view the self as a psychological interior—a private world of emotions, attitudes, wishes, secrets. The self may also be conceptualized in concrete and abstract terms. These modes of self-conceptualization will be elaborated in chapter 8.

## DIMENSIONS OF THE SELF-CONCEPT

In addition to the parts (the *content* of the self-concept) and the relationship among the parts (its *structure*), the self-concept can also be characterized in terms of its *dimensions*. An important part of the individual's cognitive structure is his system of attitudes. In the view of this work, it is both fruitful and legitimate to consider the self-concept as one of these attitudes.[7] At first glance, self-concept research and attitude research may appear to have little in common, emerging from different traditions and pursued by different practitioners. The connection, we believe, is far closer than it may appear. Every human can be characterized in terms of a large number of dispositions. Some of these, such as intelligence, optimism, and originality, are essentially object-free. But other dispositions— liking the President, disliking the university, approving of the Soviet Union, admiring movie star X—are object-bound; they reflect feelings *toward* something. The self, we suggest, is simply one of the objects toward which one has such feelings. So viewed, self-attitudes constitute part of a broader tradition of attitude and opinion investigation, enriched by its theory and utilizing its methods of research. Furthermore, object-bound attitudes can be characterized in terms of a general set of dimensions which are as applicable to the self as to any other object. Attitudes, including self-attitudes, may differ in content, direction, intensity, salience, consistency, stability, clarity, accuracy, and verifiability.

To illustrate this point, we might compare attitudes toward the self and toward State University in terms of these dimensions. (1) Content—We may ask what the individual's picture of State University is (faculty, students, location, football team) just as we may study the content of the self-picture (dispositions, social identity elements, and physical characteristics). (2) Direction—We may ask whether global attitudes toward State University are generally favorable or unfavorable, just as we may ask whether our self-attitudes are generally positive or negative, that is, whether we have high or low self-esteem. (3) Intensity—We may have strongly positive feelings toward State University, just as we may have strongly positive feelings toward ourselves. (4) Salience—To some people, State University is in the forefront of consciousness, whereas others give the matter less thought; the same is true of attitudes toward the self. (5) Consistency—We may think of State University as the ultimate in desirability but wish we were elsewhere; seemingly contradictory self-attitudes are also common. (6) Stability—Some people have firm, stable, relatively unchanging opinions about State University, whereas others' attitudes are shifting and unstable; similarly, there are those who have stable self-attitudes and others who have shifting self-attitudes. (7) Clarity—Some have a sharp, unambiguous picture of State University, whereas others may have a more blurred, vague picture; self-pictures also vary in clarity. (8) Accuracy—People's ideas or opinions about both State University and themselves may be either highly accurate or grossly false. (9) Verifiability—The average ages of State University's student body and faculty are verifiable facts, whereas the level of morale or quality of instruction cannot be so firmly established; the same is true regarding the individual's perception of his height and weight in contrast to his conception of his sensitivity or civility.

The self-concept, we suggest, is largely revealed by characterizing individuals in terms of these universal attitude dimensions. Thus, if we can learn *what* the individual sees when he looks at himself (chiefly social identity elements, dispositions, and physical characteristics); whether he has a favorable or unfavorable opinion of himself (direction); how strongly favorable or unfavorable these feelings are (intensity); whether the individual is constantly conscious of what he is saying or doing or whether he is more involved in tasks or other objects (salience); whether the elements of his self-picture are consistent or contradictory (consistency); whether his self-attitude varies from day to day or moment to moment, or whether, on the contrary, it is a firm, stable, rocklike structure (stability); whether he has a firm, definite picture of what he is like or a vague, hazy, blurred picture (clarity); whether his self-picture is correct or false (accuracy); and whether its components are susceptible of objective confirmation (verifiability)—if we can characterize the individual's self-picture in terms of each of these dimensions, then we would have a good, if still incomplete, description of the individual's self-concept.

There are, however, certain respects in which self-attitudes, although not necessarily unique, are nevertheless distinctive. One of these is to be found in the area of *importance*. Whereas attitudes toward other objects of our phenomenal fields vary widely in importance, this is not true of attitudes toward the self. To a Republican National Committeeman, the Republican Party is an important object, whereas X brand of whitewall tires may be extremely unimportant. The self, on the other hand, is an important object to everyone, usually the most important object in the world. In part, this may be attributed to the fact that the self, unlike any other object in the world, is inescapable. Our attention to other objects may shift and change, but the self is always there, always with us, and it enters into each situation with a frequency shared by no other object. In discussing the desire for fame, McDougall refers to it as "a fixed desire, because it springs from the sentiment of self-regard, and is centered upon the self, an object perpetually present, one from which it is impossible to be separated, one which is inevitably brought to mind in all situations, especially all situations that call for choice of goals and decisions as regards the means to the chosen goal" (1932).

A second distinctive quality of self-attitudes, brought sharply to the fore by Mead (1934), is that the self is a *reflexive*; the individual is both subject and object. Among all the attitudes that might be studied, then, self-attitudes are unique in that the person holding the attitude is the same entity as the object toward whom the attitude is held.

Third, the self-concept is the product of certain incommunicable information. As a consequence, the individual does not—indeed cannot—see himself as others see him. To be sure, with reference to many aspects of his social exterior (such physical characteristics or social identity elements as height, weight, hair color, sex, or race or even such behavioral characteristics as speaking rapidly or being good at basketball), the individual's view of himself and others' views of him may concur completely. But with regard to certain elements of a psychological interior, for example, certain feelings and emotions, this information is ultimately accessible to himself alone; others can learn of it only at second hand, through what the individual discloses or through the observation of overt signs. The individual knows whether he feels happy or angry, elated or depressed, buoyed up by wild fantasies or cast down by premonitions of doom in a way that no one else can.

Another reason why the concept of self cannot correspond perfectly to the concept anyone else holds of it is that it is viewed from the individual's unique point of view. Though each of us constantly makes efforts to see matters from the perspective of others, ultimately each of us is encapsulated within his own phenomenal field. We view the objects of the world, including the self, in terms of the facts of our own biography and in accordance with our own interests and concerns. In looking at the President, for example, a political reporter, a political scientist, a member of the opposition party, an uninvolved citizen, and a friend all see a "different" person; which aspects are highlighted, and what standards of judgment are employed are based on personal motive and singular experience. So it is with the self. We bring to bear upon ourselves a perspective that is unique—not necessarily more or less accurate than other perspectives, but certainly different. This special angle is made still more distinctive by the use of special lenses—the lenses of our private motivational systems. As Hobbes (1887: 63) expressed it: "For such is the nature of men, that howsoever they may acknowledge many others to be more witty, or more eloquent, or more learned, yet they will hardly believe many to be so wise as themselves; for they see their own wit at hand, and other men's at a distance."

Fourth, the general human propensity to assess and evaluate the objects which enter the phenomenal field (Osgood, *et al.*, 1957) applies fully, perhaps even particularly, to the self. Almost invariably, to see ourselves in whole or in part is to assess, evaluate, and pass judgment on what we see. We seem scarcely capable of even looking at any of our physical characteristics, dispositions, or social identity elements without immediately deploring or applauding what we observe. This obvious fact, as we shall see, has profound implications for self-esteem.

A fifth important characteristic of self-attitudes is that they give rise to a unique set of emotions, namely, pride and shame. Our emotional response to the "looking-glass self" is, in Cooley's terms, "some sort of self-feeling, such as pride or mortification" (1902: 152). This is not to say that all emotions associated with self-attitudes are entirely different from attitudes toward other objects. If we hold strongly negative attitudes toward the President and toward ourselves, the feelings toward each may well be the same —hatred and contempt. If we hold strongly positive attitudes toward a football hero and toward ourselves, similar emotions are likely to come into play, such as respect and liking. But certain emotional responses—pride or shame (or their synonyms)—are aroused *only* with regard to the self or ego-involved objects. These emotional reactions are exclusive to self-evaluations.

The term "pride" has somewhat contradictory connotations because pride can be either healthy or pathological in nature. As Helen Lynd (1958: 252) has noted: "Pride as arrogance the Greeks called *hubris*. It was overweening, a contempt for men and a defiance of the gods, which brought an inevitable tragic outcome. Very different is the Greek *philotimo*, which is honor, inviolability, freedom in oneself through selective identifications with aspects of one's own or a wider culture. If one has *philotimo*, self respect, then *hubris*, arrogance in relation to others, becomes unnecessary." In either sense, however, the term pride has reference to feelings about the self, and to that alone.

Finally, the dimensions of *accuracy* and *verifiability* have special importance in understanding the distinctive nature of the self-concept. To hold false ideas of Egyptian history and false ideas about our own characteristics may represent similar cognitive processes, but the differences in personal consequences are profound. The dimensions of accuracy and verifiability thus demand special attention.

*Truth and the Self-Concept.* Cognizant of the personal damage perpetrated by false self-concepts, wise men throughout the ages have urged their fellows to "know thyself," to shed the veil of illusion and to see themselves as they really are. Today, psychotherapy sees as a major task the need to induce the patient to recognize and accept himself for what he really is: "What we mean by insight in this context is that the self of which the person is aware comes to correspond to the inferred self—in other words, that the person comes to see himself as an informed other person sees him" (Hilgard, 1949: 381). But this view rests on the assumption that there is a verifiable self-concept which can be characterized accurately by the objective observer. We suggest that in fact the individual, even with the best will in the world, the most sincere wish to know the truth, *cannot* see himself as he really is, that he directs his life largely on the basis of unverifiable assumptions. What are the grounds for this assertion?

The most obvious reason that we do not see ourselves as we really are is that to a considerable extent *there is no way to know what we are.* As both Allport and Odbert (1936) and Gordon (1968; 1976) have made clear, the individual can view himself in terms of literally thousands of characteristics. Although impressive progress in personality measurement has been made over the years, it is still the case that, for the vast majority of traits—the ways people ordinarily think about and judge themselves—there are simply no objective scientific instruments or other evidence for determining what the individual is actually like. Where are the reliable and valid standardized instruments, scales, or measures for assessing how kind, considerate, neat, interesting, competitive, jealous, or sensitive we are? Where are the Scholastic Aptitude scores for generosity or conscientiousness, the College Entrance Examination Tests for courtesy or ruthlessness, the Stanford-Binets for cautiousness or cultural sophistication? As Eriksen and Eriksen (1972: 1) observe: "What do we really mean when we say that someone has a 'sparkling personality'? . . . Can we say that sunlight appears a hundred foot-candles brighter to any person who scores above the ninetieth percentile on the Bright-Beam Personality Sparkle Scale than to those who score lower?" Nor are the judgments of outsiders, even if consensus among them obtains, necessarily more objective or accurate than the individual's own; one may achieve intersubjectivity without achieving objectivity.

A second reason for the unverifiability of the self-concept lies in the indeterminate relationship of the disposition to its behavioral manifestation. Since a disposition is an underlying response tendency or potential, it cannot be observed directly but can only be inferred on the basis of certain overt manifestations. No one can perceive the disposition of "generosity;" one can only perceive certain actions or behavioral consequences which are used as a basis for inferring the underlying disposition.

What makes the relation between the disposition and the behavioral manifestation so often indeterminate is that the same disposition may be reflected in different behavior and different dispositions may be manifested in the same behavior. We judge someone to be very clever on the basis of some epigram he casually tosses off, but when we learn that he has simply borrowed it from Wilde or Shaw, our assessment of the underlying disposition changes. It is possible to observe the response of the other objectively; what we cannot do is to determine with certainty what the behavior, response, or evidence means, that is, what underlying disposition it reflects.

In terms of dispositions, behavior is thus susceptible to diverse plausible interpretations. Take a soldier who, in the heat of battle, rushes into the enemy stronghold; taking the enemy by surprise, he destroys or captures them. What he has done is an objective fact. But how shall his act be interpreted? Does it prove that he is a man of utmost courage, fearless to the core? Or does it mean that he is simply too stupid to recognize obvious danger when it stares him in the face? The man's act is clear, but whether the act reflects the quality of "courage" or of "foolhardiness" is a matter of interpretation. Hence, the individual who elects to place a positive construction on his act can do so without any obvious violation of reality.

So it is with most traits. On what basis shall the individual decide how "generous" he is (whether he loaned $10 to a friend), or how "punctual" (he was late to class three times last term)? There are few objective facts; their meaning is ambiguous; and they are rarely comparable from individual to individual. One person considers himself a lover of classical music because he is fond of Tschaikovsky and Verdi whereas another feels that anything less than enthusiasm for Telemann, Mahler, or Shostakovich reflects only a passion for the vulgar. Is there any way to prove that one or the other is wrong? There is thus an inherent and unavoidable ambiguity with regard to the large variety of ways potentially available to the individual for describing and assessing himself.

A third reason why it is so difficult in most cases to establish the accuracy of the self-concept is that the self-concept is largely couched in the terms of natural language. Since we see ourselves primarily in terms of the categories of thought supplied by language, the nature of language assumes critical importance for understanding our self-concepts. In this regard, however, one reason it is so difficult to describe ourselves (or anything else) dispassionately and objectively is that the words are simply not there. The fact of the matter is that the language of traits is completely shot through with evaluative overtones. Hence, Gough and Heilbrun's (1965) "Adjective Check List" includes not only neutral adjectives (descriptive terms) but 75 favorable and 75 unfavorable (evaluative) terms. Even a superficial glance at any list of adjectives shows that the vast majority are not simply value-free descriptions but imply negative or positive judgments. To call a person "kind" is not a description; it is an accolade. To call him "cruel" is not to describe him, but to condemn him. If we say someone is "ugly," we are providing a much better description of *our* feelings than of *his* features. The work of Osgood *et al.* (1957) has documented empirically the overwhelming importance of the evaluational component in judgments of all objects, including self-judgments. In most respects humans appear scarcely capable of seeing something without at once passing judgment on what they perceive; and this propensity is entrenched in their language. The words we use to describe something express not simply what we perceive but also how we feel about it.

This peculiar characteristic of a great deal of trait language produces the anomaly that *opposite* linguistic expressions become entirely appropriate to the description of the *same* behavior. For example, if one person calls us clever and thoughtful whereas another calls us shrewd and calculating, are they really describing anything different? A person who is aggressive and ruthless considers this behavior to mean that he is fearless, courageous, and free of false sentimentality. A person who is seen as compliant and unwilling to stand up for his rights interprets this behavior to mean that he is an agreeable person, eager to help, unselfish and willing to sacrifice for others. Each of us will agree that "I am firm, thou art stubborn, he is obstinate."

Such terminological selectivity applies to statuses as well as to traits. Accompanying the process of social labeling is an active process of re-labeling designed to strip the label of its pejorative connotations. Insane asylums become mental hospitals, the deaf become hard of hearing, Negroes become blacks, Indians become native Americans, queers become gays, bastards become illegitimate, and, when this term loses its moral neutrality, out-of-wedlock children (more recently OW's), garbage collectors become sanitary engineers, bookies become turf accountants, and so on. Thus the nouns of social identity are generally as prone to social evaluation as the adjectives.

Two modes of re-labeling are especially worthy of mention. The first is the absorption of an abjured category into a respected one. While this appears to be a mere sleight of hand, it actually results from the fact that the same term may have different dictionary meanings. A striking case in point—and one with which one can easily sympathize because it aims to protect the self-esteem of children—is the term "exceptional children." One of Webster's meanings of exception corresponds to its meaning to most of us, namely, "better than average; superior." But another meaning is "forming an exception; rare." Child psychologists and educators have elected to interpret these words in a statistical sense, thus including those incapacitated in some way under this rubric since they are, statistically speaking, exceptional.[8]

The second form of re-labeling is to cast a socially-abjured category into the morally-neutral category of scientific terminology. It is the "medicalization of deviance" which has so changed society that there are no longer crazy people, only sick ones. Drug addiction and alcoholism, previously terms of condemnation (reflected in the popular terms "junkie," "hophead," or "souse") have been redefined as medical problems: the individuals involved are more to be pitied (or helped) than censured.

This is not to suggest that *nothing* about the self is verifiable. For the great majority of social identity elements (sex, age, race, nationality, family status) and physical characteristics (height, weight, hair color), the facts are available and their accuracy can be assessed. Indeed, substantial inaccuracy in any of these regards (the oldster who believes he is an adolescent, the five-footer who thinks he tops six feet, the trash collector who believes he is Napoleon) is considered virtually decisive evidence of severe mental disorder. In addition, good instruments have been developed for the measurement of certain types of competence and a number of "personality traits." But for a large proportion of self-components, no unequivocal evidence about the reality exists. It is not simply that people seek to avoid the unpalatable truth about themselves, although they assuredly do, but that even when they search desperately for the truth, they often cannot find it.

It is sometimes said that it is extremely difficult to study the self-concept, to learn what the individual thinks of himself and how he sees himself. That is true. But it is infinitely more difficult, we believe, to learn what the individual *actually is*.

*Self-Confidence and Individuality.* Although these two concepts lack exact counterparts in attitude theory, they are nevertheless important dimensions of the self-concept. The connection between self-confidence and self-esteem is obviously a close one, and the two concepts are often used interchangeably. Yet a good deal of confusion has arisen as a result of failure to recognize that a distinction exists. Self-confidence essentially refers to the anticipation of successfully mastering challenges or overcoming obstacles or, more generally, to the belief that one can make things happen in accord with inner wishes. Self-esteem, on the other hand, implies self-acceptance, self-respect, feelings of self-worth. A person with high self-esteem is fundamentally satisfied with the type of person he is, yet he may acknowledge his faults while hoping to overcome them.

A good deal of important theoretical and empirical work on the self-confidence dimensions is currently in progress. Franks and Marolla (1976) describe it as "inner self-esteem" (differentiating it from an "outer self-esteem", presumably based on reflected appraisals), Brim (1974) refers to it as "personal control," Smith (1968) as "competence," and many people as "internal locus of control." The theme common to all these ideas is that of feeling oneself to be an active agent in one's own life (rather than the object of external forces), of believing generally that one can work one's will on a more or less recalcitrant world.

Since self-esteem, in the sense used here, refers to a positive or negative evaluation of the self, self-confidence may contribute to self-esteem but is not identical with it. A major reason is that some people do not stake themselves on competence and mastery. To them being lovable, moral, self-sacrificing and helpful are major concerns; they may be quite contented to leave the mastery of life's harsh problems and challenges to others. On the other hand, there are those abundantly endowed with ability and talent who are confident of their ability to succeed in most tasks but who lack self-respect because they cannot be first in everything, cannot command the love of another, or are overwhelmed by a denigrated social identity element.

It is important to keep in mind the distinction between self-confidence and self-esteem since many apparently contradictory results are reconcilable when viewed in this framework. Consider, for example, the oft-repeated assertion that sexist discrimination and socialization to feminine roles produce damaged self-concepts in girls. Some studies appear to support this assertion, others to challenge it. In an exhaustive analysis of the empirical literature, Maccoby and Jacklin (1974) suggested that girls may not have lower self-esteem but do appear to have lower self-confidence. Similarly, even if disprivileged groups were found to have self-esteem levels equal to those of the more advantaged, an investigator would still wish to learn whether they might have lesser self-confidence in mastering the problems of the world, given the more forbidding societal barriers they face.

The final dimension of the self-concept, not precisely analogous to other attitude dimensions, is individuality. It was one of Durkheim's signal theoretical contributions to show how social factors heighten or blur the development of a sense of individuality among people. In his earliest work, Durkheim (1947) pointed out that societies achieve solidarity essentially in two ways. The first type of solidarity—mechanical solidarity—is based upon the similarity of members of society. In simple societies, for example, people follow the same customs, engage in the same occupations, hold the same conceptions of right and wrong, worship the same totems—in general, adhere to a common and relatively unchanging tradition. This similarity binding them so closely together simultaneously dampens their awareness of their uniqueness and differentiation from others.

The second type of social solidarity—organic solidarity—derives from interdependence based on differences. In the division of labor, each person is dependent upon the special skills of others. Each person, says Durkheim, "has a sphere of action which is peculiar to him, i.e., a personality. . . . The activity of each is as much more personal as it is more specialized" (131). A society that makes us aware of how we differ from others heightens our sense of individuality.

Contemporary American society, characterized by a refined division of labor, presents structural conditions conducive to the development of individuality. So deeply ingrained is this idea in our society that we are apt to overlook the fact that it is a fairly recent historical development, emerging during the Renaissance, receiving strong ideological impetus from the Protestant Reformation, and taking shape through an Industrial Revolution based on a refined division of labor. Despite recent writings on conformity, the idea of individuality has probably reached its zenith in contemporary American society, which has developed a virtual cult of individual personality. A society, such as American society, which incorporates into its institutions and value system the principle that all people are worthy of respect, that their individual wants and interests must be considered, that the maximum autonomy consistent with social control be permitted, that self-realization is desirable, and that, even in religious matters, the individual's relationship to his God should be a personal one—such a society is likely to foster a high level of individuality. As Williams (1951: 437–8) says: ". . . it is typical of the culture that the question as to whether there is actually such an entity as 'the individual,' 'self,' or 'ego' is usually not even thought of, and, if raised, is greeted with surprise or shock. *Of course* individuals exist, of course they have separate individual needs and rights."

To summarize this discussion of self-concept dimensions, we have suggested that there are certain factors especially characteristic of, or completely unique to, self-attitudes. Unlike attitudes toward other objects, the self is important to everyone; the person holding the attitude and the object toward whom it is held are the same; it is judged and evaluated; there are certain kinds of emotional reactions (pride and shame) which are unique to the self and ego-involved objects; and accuracy is difficult to ascertain because of low verifiability. At the same time, all the fundamental dimensions of attitudes are completely relevant and significant aspects of the self-concept. No characterization of the self-concept can claim to be adequate which fails to take account of these various dimensions.

Most of the dimensions of the self-concept, it should be pointed out, are relevant both to the parts and to the whole. A person may have positive or negative feelings toward himself as a whole, and he may also have positive or negative feelings toward his various traits—his intelligence, helpfulness, athletic skill, and so on. The same would hold true of such dimensions as stability, clarity, and self-confidence. Similarly, salience may refer either to specific components or to the whole. When we speak of specific salience, we mean that certain components of the self stand out, are in the forefront of attention. Mulford and Salisbury (1964), for example, show that, to the woman, the status of mother is more salient than the status of father is to the man. McGuire *et al.* (1978) provide evidence that ethnic identity is more salient when the individual is surrounded by those of a different ethnic group. Global salience, on the other hand, refers to the degree to which the self is prominent in the individual's total phenomenal field. In taking a test, which component dominates one's thoughts: the task at hand or the question of how well one is performing? Part of the complexity of the self-concept thus rests in the fact that the various dimensions apply equally to the whole and to the parts, that both are important, and that one cannot generalize from one to the other.

## EGO-EXTENSIONS

The final area of the extant self-concept is the individual's ego-extensions. Ordinarily we think of ourselves as bounded by our skins: there we start and there we end, and ever it shall be. And yet each of us knows that the self stretches out to encompass elements external to it. We at once recognize the independent identity of these external things, people, or groups but at the same time feel they are part of us—indeed, that in a sense they *are* us. William James explicitly included these external components in what he called the "empirical self":

> The Empirical Self of each of us is all that he is tempted to call by the name of *me*. But it is clear that between what a man calls *me* and what he simply calls *mine* the line is difficult to draw. We feel and act about certain things that are ours very much as we feel and act about ourselves. Our fame, our children, the work of our hands, may be as dear to us as our bodies are, and arouse the same feelings and the same acts of reprisal if attacked. . . .
>
> In its *widest possible sense*, however, *a man's Self is the sum total of all that he CAN call his*, not only his body and his psychic powers, but his clothes and his house, his wife and children, his ancestors and friends, his reputation and works, his land and horses, and yacht and bank-account. All these things give him the same emotions. If they wax and prosper, he feels triumphant; if they dwindle and die away, he feels cast down—not necessarily in the same degree for each thing but in much the same way for all. [1890: 291–2]

The extant self-concept thus includes the individual's "ego-extensions," for they are experienced as a part of what "we" are. But where do we draw the line? Where do we cross the border from self to nonself? Where does the "I" end and "non-I" begin? One must agree with James that the division is rather vague and nebulous, that the boundary is blurry. My child, my father, even my car may be very much a part of myself, but is the Governor of my State also a part of myself?

How, then, shall one decide whether or not an object or element is a part of the self? There are, we believe, several distinguishing features of what Cooley has felicitously called the "appropriation" of exterior objects by the self.

The first is the subjective experience of "me" or "mine." Is it *the* university or *my* university, *the* President or *my* President, *the* company in which I work or *my* company? The incorporation of the "external" element into the self is exclusively a matter of subjective experience. The mother who speaks of "my" child or the author who speaks of "my" book, though referring to objects physically separate from themselves, may be speaking of something central to their feeling of self.

It thus follows that the self does not have fixed and rigid boundaries. It can take into itself more and more objects and individuals, more and more external things whose fate then becomes wrapped up in its own. But the self can also contract, sloughing off elements which are then experienced as detached from the self. A childhood friend, so intimate a part of the self in early years, is thrust to the periphery of the self in adulthood. The separation of time and space similarly make our siblings, cousins, uncles, and aunts less central components of the self. The boundaries of the self are more like a balloon than a solid sphere; the self may shrink and stretch.

The second defining characteristic of ego-extensions is the presence of the emotions of pride or shame. As noted in our earlier discussion, pride and shame, unlike other attitude dimensions, have reference *only* to the self. I may admire Beethoven, but I take no pride in him; I may despise Hitler, but I do not feel ashamed of him. But if anything external to the self is capable of arousing feelings of pride or shame—pride in my shiny new automobile, shame at my unfashionable clothes, pride in an honor bestowed, shame or embarrassment at the defeat of my school team—then these elements have been appropriated by the self and are contained within its boundaries.

A third distinguishing characteristic of an ego-extension is the phenomenon of *introjection*, defined by Webster as "the adoption of externals (persons or objects) into the self, so as to have a sense of oneness with them and to feel personally affected by what happens to them." When an external object is introjected, the fate of the object and of the self are experienced as inextricably intertwined. The success of these externals inflates self-esteem, their failure diminishes it. The degree to which we take praise or blame of this object *personally* may help to define the boundary of the self. Similarly, the "family skeleton" reflects the needs of family members to conceal from public view the peccadilloes of one of its members at the risk of shame to all.

One reason it is so difficult to define the self-boundaries sharply is that the introjection of external elements may differ in degree or intensity. In a study of school children in Baltimore (to be discussed in later chapters), subjects were asked: "If someone said something bad about your mother, would you almost feel as if they had said something bad about you?" The same question was asked about a number of other possible ego-extensions. These ego-extensions might be other people (father, brothers and sisters, friends, President of the United States, governor of your state); groups (race, religion); material objects (clothes, toys or hobby equipment); institutions (school); geographical region (neighborhood); ego-extension's own statuses (father's job); and "work" (school work). Although all are experienced as ego-extensions, some are central to the self, others peripheral. For example, whereas 89 percent would take personally an attack on the mother and 82 percent an attack on the father, only 67 percent would so interpret an attack on their school work, 50 percent an attack on their schools, 32 percent an attack on their toys or hobby equipment, and 24 percent an attack on their state governors. Whether external objects are experienced as a part of the self may be more a question of degree than of kind.

Few external objects are experienced as so central to the self as those representing the outcome of our own efforts.[9] The artist's painting, the author's book, the furniture maker's chair, the shop-owner's store are felt to be a part of their selves. To tell an author: "That's an asinine book, but don't take it personally; it's the book I'm criticizing, not you" is to provide cold comfort to the writer. Similarly, a house we have built, a tree we have planted, a garden we have designed are felt to be much more intimately "ours" than if we had paid someone else to do it. Even the indirect and mediated products of our efforts can stimulate feelings of pride or shame. A mother may take pride if her child achieves professional success, an athletic coach if his pupil wins an Olympic victory, a professor if his former assistant authors a successful book because the achievements of these ego-extensions are felt to represent the outcomes of one's own efforts and energies. Even money, cold and impersonal as it may be, may nevertheless be an important ego-extension insofar as it symbolizes our own successful efforts. Although inherited money commands exactly the same goods and services in the marketplace as earned money, it nevertheless has a different psychological significance. Similarly, people who descend into deep depression or even commit suicide when their businesses go bankrupt do so not simply because of the material deprivation anticipated but because the sense of self is diminished and partially negated by the loss.

It is noteworthy, from a developmental viewpoint, that the propensity to incorporate external elements into the self appears in the early stages of self-concept development; the distinction between "me" and "we" may actually be more blurred at this time than later. Expressions of hostility among children often take the form of attacks on the other's ego-extensions, with the clear-cut awareness on the part of both the attackers and the attacked that the deprecation of the ego-extension serves as a deprecation of the self. Various taunts and imprecations, such as "your mother is a junkie" or "your father is a pimp" are plainly intended as an attack not on the parents but on the child, *and are plainly recognized as such by both participants in the interaction.* The catcall: "Shakespeare is a junkie" would hardly elicit anything but an openmouthed stare. That we do not outgrow this propensity, however, is reflected in Hobbes' (1887) observation that "men are brought to battle for any sign of under-value, either direct in their persons, or by reflection in their kindred, their friends, their nation, their profession, or their name."

It is important to clarify the connection between ego-extensions, as used here, and Sherif and Cantril's (1947) ego-involvements. In a critique of ego-involvements, Smith (1950: 520) noted that "widely divergent sorts of psychological facts turn out to be embraced by the term . . . more often than not ego-involvement means the involvement of a person's pride and self-esteem in the task; he feels put to the test and ready to be ashamed of a poor performance. In other instances the term is invoked to cover immersion in a cause or falling in love—cases in which the person, to be sure, cares as deeply about outcomes as in the first type but may be engrossed to the point of losing awareness." In our usage, only those elements felt to *reflect upon the self* constitute ego-extensions.

These, then, are the chief components of the *extant* self-concept: content, structure, dimensions, and ego-extensions. Together they constitute most of what the individual sees and feels when he reflects upon himself at a given point in time.

## Desired Self-Concept

Alongside of, and frequently intertwined with, our pictures of what we are like (the extant self) are our pictures of what we wish to be like (the desired self). The ability and even the propensity to imagine ourselves as other than we are is a remarkable feature of the human mind, having profound emotional and behavioral consequences for the individual. To a large extent, the extant self-concept is viewed against the backdrop of the desired self and acquires a distinctive meaning in that setting. As a trivial example, two people who consider themselves mediocre tennis players will have radically different attitudes toward their level of performance if one is entirely contented with that level whereas the other strongly wishes to be outstanding; self-satisfaction or self-hatred may thus be the responses to the identical extant self-concept. The desired self-concept is almost invariably the reference point against which the extant self is viewed and judged.

Like the extant self-concept, the desired self is not all of one piece. Three aspects or components of the desired self may be distinguished: the idealized image, the committed image, and the moral image. Although each of these can also be characterized in terms of self-concept content and structure, in this discussion we shall simply present the main features of each type of desired self.

*Idealized Image.* The concept of the idealized image was introduced into psychoanalysis by Karen Horney in *Our Inner Conflicts* (1945), subsequently becoming the cornerstone of her psychoanalytic theory of neurosis. The child, she held, is born with a certain potential but requires favorable environmental conditions in order to realize this potential. Just as an acorn possesses the potential to grow into an oak but will not do so unless certain conditions of soil, water, and sunshine are present, so the individual is born with certain human potentials which require favorable conditions for realization. In the face of unfortunate childhood conditions, however—a parent who is neurotic, egocentric, self-centered, overdemanding, inconsistent—the child develops *eine Urangst*—a "basic anxiety." Seeking to overcome this intolerable fear, the child attempts to gain security by lifting himself above others. He creates in his imagination a picture of someone who is the ultimate in beauty, intelligence, kindness, goodness, or whatever. This imaginative construct comes to be the type of person he would like to see himself as. It may be a picture of a person who is always kind, cheerful, and popular; a perfect housewife and mother; a creative, inspired genius; a hard-driving, ruthless, successful businessman; a detached, serene, contemplative philosopher; and so on. Sometimes the image is a cultural stereotype—a Florence Nightingale, a Napoleon, a knight in shining armor, an Albert Schweitzer. More often it is a combination of components from various sources.

In Horney's view, idealization is the comprehensive neurotic solution, promising to solve the problems of feeling lost, anxious, or inferior by providing the individual with a "much-needed feeling of significance and superiority over others." In the neurotic, this idealized self comes to replace the real self as the focus of interest, dominating the individual's thoughts, feelings, and actions.

As a clinician, Horney was keenly sensitive to the pathological nature and consequences of the idealization process. First, the individual feels *driven*—he *must* make the most money, have the most dates, get the best marks, be the best hostess. Second, neurotic idealization is *indiscriminate*. Although most people want to be liked, the neurotic wants to be liked by *everyone*. Third, this image is characterized by the quality of *insatiability*. However much prestige, money, or conquests, it is never enough. Fourth, the neurotic's *reactions to frustration* are inordinate. While everyone experiences disappointment or anxiety, the neurotic's reaction may give rise to panic, despair, and rage entirely out of proportion to the occasion.

Several further consequences issue from this preoccupation with the idealized image. One is a feeling of intense *strain*, since it is always difficult to live up to such high standards. A second is *hypersensitivity to criticism*, because even a hint of criticism from others challenges the idealized image, causing the neurotic to feel depressed and angry at himself and others. A third is extreme *vulnerability*, since the individual is easily and deeply hurt if he fails to live up to his standards or otherwise falls short of his idealized image.

The ultimate result of all this, in Horney's view, is self-hatred and self-contempt. The idealized, glorified self becomes a "measuring rod with which to measure his actual being. And this actual being is such an embarrassing sight when viewed from the perspective of a godlike perfection that he cannot but despise it" (1945: 110).

Because Horney, as a clinician, developed her theory largely on the basis of psychoanalytic practice with neurotics, one might assume that it has little explanatory value for the psychological processes of normal people. This assumption is far from true. *Every* human being has the gift and curse of imagination, the uniquely human power to imagine himself as other than he is. What differentiates the neurotic from the normal, then, is not the creation of an idealized image but the pathological consequences of neurotic idealization. The fact that the neurotic uses his idealized image as an instrument for overcoming basic anxiety while the normal person does not should not be taken to imply that the latter lacks this imaginative construct. The idealized image is an important component of everyone's self-concept.

*Committed Image.* Were each of us to judge himself by the standards of the idealized image, feelings of inadequacy would be universal. But the desired self contains another imaginative product —a committed image. This self-picture is one that we take seriously, not simply one that is pleasurable to contemplate. Everyone has dreamed of himself as other than he is, has savored in his mind a pleasing self-picture—as football hero, movie star, brilliant surgeon, dazzling political orator, perfect hostess, or artistic genius. Particularly in youth, when the world is rich with possibilities and the fantasy life is vivid, such Walter Mitty dreams are common. In real life, of course, most people end up driving trucks, tending lathes, selling clothing, caring for children, operating computers. Part of the reason they are not drowned in self-contempt is to be found in the fact that they compare their achievement with their committed, not their idealized, images—committed images which may well be within their reach.

Both the committed and the idealized images, then, constitute parts of the desired self. Much of human striving is based on the individual's efforts to convert himself into one of the pictures. Neither the idealized nor the committed image is the picture of how the individual sees himself but of how he would *like* to see himself (although among neurotics the distinction is frequently blurred). Neither is a literal photograph capturing what is real. Both are portrait paintings, reflecting not reality but visions in the mind of the artist.[10] But whereas one is a realistic picture of self that an individual has committed himself to, the other is a glorified vision, warts excised, flaws vanished, virtues triumphant. Implicitly, at least, most people recognize the distinction within themselves. It is only when the distinction becomes blurred or when the idealized image dominates the individual's thought to the exclusion of all else that psychopathology appears.

The distinction between the idealized and the committed image is frequently overlooked in the literature. One of the best-known measures in the self-concept literature is the "self-ideal" discrepancy measure (Butler and Haigh, 1954). Subjects are asked to describe themselves as they actually are in terms of a large number of statements and then to describe themselves as they would like to be "ideally" in terms of these statements. But how does the respondent interpret "ideally"—as a self unrestricted by any bounds of reality, little more than a playful fantasy, or as the self he earnestly wishes to become, or as something in-between? The meaning of "self-ideal" discrepancy scores will be entirely different, depending on the interpretation.

It is noteworthy that the difference between the idealized and committed images is recognized even by very young children. Piaget (1928: 246–71) has observed that children as young as 3 already are successful in distinguishing "between two kinds of existence—what is true and what is simply imagined. . . . From this date onwards children distinguish better and better ideas that are believed *pour de vrai*, as the young Genevans say, and ideas that are believed *pour s'amuser*." The Baltimore study, to be discussed later, which included children as young as third-graders, asked subjects: "You have been telling me what you might want to be like when you grow up. But what if you COULD BE ANYTHING IN THE WHOLE WORLD, anything at all, what would be the *most wonderful thing* you'd want to be when you grow up?" After the interviewer recorded the most wonderful things, he then selected the first and asked: "Would you *really* like to be _____ or is it just a *nice idea?*" It was apparent from the earnest head-nodding of those who said they really did, or the rather abashed smiles of those who said it was just a nice idea, that even young children clearly recognized the distinction between a playful fantasy and a serious commitment to a desired self.

An interesting example of this distinction was reported in one of the earliest works of empirical social research (Jahoda, Lazarsfeld and Zeisel, 1971), a study of an unemployed village in Austria in the early 1930s. When a little boy was asked what he would like to be when he grew up, he replied that he would like to be an Indian chief, but, after a pause, added wistfully: "But I don't know if I can get the job."

*Moral Image.* Interlaced with the issue of what we *wish* to be is that of what we feel we must, ought, or *should* be. For lack of a more satisfactory term, we will call it the "moral image," despite its inexact connotations. Each person is guided by an implicit book of rules that he feels he must follow, a set of standards he feels obligated to meet. This process is immanent in the development of the self as object. In the words of Simmel (1950: 99): "Morality develops in the individual through a second subject that confronts him in himself. By means of the same split through which the ego says to itself 'I am'—confronting itself, as a knowing subject, with itself as a known object—it also says to itself 'I ought to.' The relation of two subjects that appears as an imperative is repeated within the individual himself by virtue of the fundamental capacity of our mind to place itself in contrast to itself, and to view and treat itself as if it were somebody else."

In this process we can observe an almost pristine expression of what the existentialists refer to as "self-objectification" (Tiryakian, 1968)—the individual standing outside himself, viewing himself as an object, passing judgment on what he sees, molding the object into a certain shape. We are aware of this bifurcation of the self when we refer to someone as his "own worst taskmaster" or as a person who "pushes himself" too hard. The imagery of someone "pushing himself" is as curious as it is apropos: the individual driving or shaping the self in accord with what he believes it should be.

The moral image is thus a set of standards, a system of shoulds (and, of course, should nots). The content of this system of shoulds derives partly from social indoctrination, partly from idiosyncratic selection. As understood here, the moral image is broader than the conventional notion of conscience, for it includes the entire structure of self-demands, the total system of standards and requirements that the individual imposes on himself. Although there is considerable overlap, it is convenient to differentiate three aspects of the moral image: conscience, role demands, and idiosyncratic self-demands.

The standards internalized in the conscience (or superego) are primarily those associated with the conventional norms of morality and are based largely on prohibitions to action: one should not lie, steal, covet, fall prey to alcohol, drugs, sloth, self-indulgence, and so on. These rules of conscience are taught and reinforced by such social institutions as the family, school, and church. Violation of these norms may give rise to severe pangs of guilt or self-hatred.

The structure of role demands is a set of prescriptions and proscriptions associated with behavior required of, or prohibited to, status incumbents. Thus, the nurse feels she should be kindly and sympathetic; the doctor, dedicated; the man, forceful and aggressive; the mother, patient and loving; and so on. Much of human behavior conforms to these demands.

Finally, in the course of life people develop a system of personal and idiosyncratic self-demands which are no less imperative than the social demands. One person feels he should not need more than eight hours of sleep and condemns himself if that does not suffice. Another feels he should always be cheerful and good-humored, reproaching himself if he is in a bad mood. A person who feels he should always be rational is appalled when he commits a logical blunder, even though he might tolerate such an error in another as a simple human failing. A writer believes that he should produce a perfect work in a burst of inspiration without revision, additional drafts, or further work. A student feels he should get an A on every test and becomes enraged at himself if he does not. Someone feels he should be able to work full-time, carry a full academic course load, engage in extracurricular activities, and have an active social life—all without strain or fatigue.

This complex of shoulds and should nots may be the highest product of human reason or the outcome of the grossest superstition. Allport (1955) has distinguished the primitive or immature conscience, which is based on the "must," from the mature conscience, which is based on the "ought." The former is particularly characteristic of children, who accept the fact that certain behavior is right or wrong because it is based on the unchallengeable authority of the adult. The latter is characteristic of the mature intellect, the person who understands the human and social desirability of loyalty, brotherhood, tolerance, honesty, and so on and thus has a rational basis for the system of moral rules he has adopted.

The special significance of the moral image—whether musts, oughts, or shoulds, whether socially structured or individually selected—is that its violation is attended by *self-condemnation*. Insofar as matters of conscience are concerned, such violation is experienced as shame or guilt. The power of guilt or shame has long been recognized, and their motive force has usually been considered in association with the rewards and punishments of earlier years. It seems apparent, however, that these experiences represent direct blows to current self-esteem. The soldier who flees under fire, or the pillar of respectability revealed as an embezzler or a pervert may choose suicide as an alternative preferable to facing both society and himself. But failure to live up to the moral image may also have more pervasive, if less dramatic, manifestations. A dieter may experience stronger pangs of guilt, and consequent feelings of self-hatred, at consuming a hot fudge sundae than at violating the sex mores or cheating an insurance company.

The individual is thus subjected to an elaborate structure of self-demands to which he submits, or attempts to submit, on pain of self-condemnation. The question is not whether people are *influenced* by these self-demands but whether, in Horney's words, they are *tyrannized* by them. Indeed, one important but neglected area of research is the individual's tolerance or intolerance of his failings. Many people who have a sympathetic and understanding attitude toward the natural human failings of others have no such tolerance for themselves. They understand that others may become tired, make mistakes, or yield to biological drives, but they flog themselves mercilessly for failing to live up to their self-demands.

The gap between the extant self-concept and the desired self (idealized, committed, moral) is thus a source of perpetual concern to the human being. One of the reasons for consigning events or feelings to the unconscious is the fear of having to see a self that we cannot stand—a self too different from the desired self to bear. The fear of admitting to oneself that one feels greed, lust, hatred of parents, homosexual desires, sadistic inclinations is the fear of looking at a self that is intolerable. This is the situation of the self-concept deploring the self, the "me" finding the "I" too horrible to bear, the superego condemning the id.

Two general points about the desired self are worthy of attention. The first is that the desired self is a motive force, a spur to action. It is not, like so much of human knowledge, an inert mass of thoughts lodged in the mind, waiting to be accessed when needed.

On the contrary, the individual is constantly at work seeking to shape himself into this imaginative vision. He works to become a "better person", strives to realize his ambitions, struggles to censor his unacceptable impulses, selectively directs his efforts in the immediate situation or in the long run to converting the self he observes into the self he wishes to observe or to maintaining a desired self against the forces that threaten to blemish it. It does not seem extravagant to assert that much of human behavior is motivated by the wish to attain, to maintain, or to retain a desired self.

The second point is that, although it is meaningful to distinguish among an extant self, idealized self, committed self, and moral self, the individual is not always successful in compartmentalizing these images. We may confuse what we wish to be, or feel we should be, with what we are, arrogating to ourselves the glorious qualities observed in our fantasies and resenting society for not according us the respect and deference due such a wondrous creature. We may flog the actual self unmercifully for falling short of the stern array of "shoulds" which serve as its standards of worth. An understanding of the self-concept thus requires an understanding of the *relationship* among these various self-concept regions.

In sum, it is plain that we can never attain an adequate grasp of the self-concept without taking account of man's extraordinary tendency to visualize himself as other than what he is, to construct in imagination a picture of what he wishes to be. No one can see himself as bad or good, admirable or contemptible, except with reference to the standards he has set for himself. Hence, the individual's idealized image, committed image, and moral image must all be considered parts of "the individual's thoughts and feelings with reference to himself as an object"—that is, of the self-concept.

## The Presenting Self

Implicitly and explicitly, we are always engaged in "impression management" (Goffman, 1956), governing, guiding, and controlling our own actions, acting in accordance with the type of person we wish to appear.[11] Whatever social ingredients are used in the formulation of the role, we are both script authors and central actors. No actor can portray Hamlet without a defined image of the character and an effort on his part to behave in a fashion overtly portraying it. And just as the self portrayed on the dramaturgical stage may or may not correspond to the innate propensities and inclinations of the performers, so does the self we intentionally present on the social stage match (or diverge) from the self-pictures we inwardly hold.

The form and content of the presenting self is likely to go beyond specific traits to include some of the broader social *types* described earlier. A person may seek to present himself as an average Joe, a devoted father, a dedicated scientist, a dynamic leader, a free spirit, and so on. These types will vary from thinly constructed dramaturgical roles—consisting of little more than a single trait— to fully fleshed out characters with clearly defined statuses, attitudes, values, actions, and goals.

This presenting self, it should be stressed, is not the same in all situations. As James (1890) noted, we present one side of ourselves to our family, another to our clubmates, a third to our employees, a fourth to our students. Many a youth, he notes, demure enough before his parents, swears and swaggers like a young pirate when in the company of his tough young friends. Similarly, Gordon Allport (1961) points out that the flinty-eyed businessman, unbending as steel in dealing with business associates or employees, may be soft as a marshmallow in dealing with his pretty young daughter.

Given such sharply varying behavior, some theorists have doubted whether there actually is any such thing as a self-concept. Their argument is that if people allegedly behave in a fashion consistent with their self-concepts, then what sort of will-o'-the-wisp entity is it that is responsible for such widely varying behavior in diverse situations? The answer is that, although this presenting self is much more variable and situation-bound than other self-concept components, it is by no means chameleon-like. At the core, and frequently cross-situationally, there is a more or less consistent self that we elect to present to the world. Certain of its features will come to the fore in one type of situation, other features in another, but a more or less consistent, if complex, presenting self is also characteristic of human beings. We generally want other people to think of us as a certain type of person, and make efforts to insure that they do.

## PRESENTING SELF MOTIVES

Why do people "put on an act" rather than simply "be themselves"? Although some writers attribute this propensity to individual psychopathology, the disposition is more fundamental and characterizes the healthy personality as well as the ill. There are several objectives in presenting certain selves: (1) the fulfillment of ends, goals, or values; (2) the self-consistency and self-esteem motives; and (3) the internalization of social roles.

*Means and Ends.* First, the presenting self is the chief instrument available to the individual for the fulfillment of those personal goals, aims, or ends which implicate other people. These goals may be as specific as selling a product to a customer or as broad as establishing a general reputation for morality. Does someone want a job? He must consider how to act in order to impress a prospective employer. Does he wish his love to be requited? He must think about the appearance and behavior that will arouse feelings of love in the object of his desire. Does he seek membership in a club? He must decide whether dignified or casual, conformist or nonconformist, emotional or restrained behavior will elicit the desired response.

From a sociological viewpoint, it is important to note that the nature and prevalence of such presenting selves may be strongly influenced by social and historical factors. This is vividly illustrated in Fromm's (1947) discussion of the "marketing personality." In the earlier stages of a capitalist economy, Fromm pointed out, the focus was on production. In a competitive market economy, the most efficient producer prospered, the least efficient failed. As technology developed, however, production became standardized, thus eliminating the productive advantage of one competitor over another. The realm of competition thereupon shifted to sales. Since technology insured that the products and prices of all producers were more or less interchangeable, the chief bases for gaining customer preference were more attractive packaging, more effective sales claims, more successful types of "impression management." A salesman, no longer able to sell his products on the basis of superior price and quality, had to induce his customers to buy because they liked *him*— to sell his "personality," not his product. According to Fromm, this orientation currently dominates social life: a candidate is elected because he is likeable, a minister is chosen because of his dignified manner, a girl is married because of a "pleasing personality," a doctor is selected because of his "bedside manner," and an executive is promoted because of his social skills. In all these cases, success involves the presentation of a certain type of self.

But broader and more fundamental objectives also underly presenting selves. One of the major aims or ends of human striving is social approval. Hence, the tremendous force for conformity in society or its subgroups. The high-schooler who wants to be accepted by a "fast crowd" must dress a certain way, act a certain way, approve of certain things, violate certain rules, and so on. The delinquent who aspires to acceptance in a delinquent subculture is just as rigidly bound by an elaborate system of behavioral and attitudinal prescriptions. If we wish to command the favorable opinions of different types of people, we may be obliged to present different selves to different types, depending on *their* attitudes and values.

*Self-consistency and Self-esteem.* A second objective in presenting a certain self is the satisfaction of the self-consistency and self-esteem motives. As we shall see in the next chapter, everyone has a need to establish and protect a self-picture. Hence, selves are presented in order to *confirm a self-picture* (if it is established) or to *test a self-hypothesis* (if it is in process of formation). Although, in both cases, the immediate aim may be to convince others that we are a certain type of person, this objective serves the deeper purpose of convincing ourselves. However much an individual may like to think of himself as intelligent, attractive or skilled in some way, in the long run he will actually believe he is so only if this view is substantiated by external evidence. One of the major sources of evidence is the responses of others toward us (consensual validation). It is difficult to maintain our self-view as a profound philosopher if others consider our ideas trite or trivial; as a wit, if they yawn at our jokes; as a dignified figure, if they puncture our pretensions.

For youth in particular, self-presentation is likely to assume the function of testing, and attempting to validate, one or more self-hypotheses. The adolescent may thus reflect on many different possible selves, playing diverse roles, cultivating diverse abilities, and experimenting with different traits. In a groping and tentative way, different selves may be rehearsed—the glamor girl, the caustic wit, the world-weary sophisticate, the dedicated revolutionary. Selves may be tried on or discarded like garments as adolescents attempt to convince others of their sophistication, their cheerfulness, their allure, their intelligence. Hence, the common sense question: "What is he trying to prove?" has profound psychological significance, for that is precisely what the adolescent is trying to do. A girl wants a date in order to have *evidence* that she is attractive or loveable. A boy wants to succeed in school in order to gain *evidence* that he is intelligent. What concerns these youngsters is the *uncertainty* about what they are like; what motivates them is the desire to know, to achieve certainty. When an adolescent—or anyone else —tries to achieve a certain goal, he does so not simply for the direct advantage it affords, but because it enables him to *prove* something about himself to himself. Yet ultimate certainty forever eludes us so that the responses of others are required not only for confirmation but for the lifelong reconfirmation of our working self-hypotheses.

Another major objective in presenting certain selves is to protect and enhance self-esteem. Since the perceived self—what we believe others think of us—and our own self-esteem are closely associated, people are at considerable pains to insure that others think well of them. For the cocktail party *bon-vivant* who attempts to dazzle the assembly with his wit, the adolescent striving mightily to entertain his date, the conspicuous consumer attempting to impress others with his pecuniary strength, the motivation is the same—to act publicly in such a way as to make a favorable impression on the minds of others, in order thereby to produce a favorable impression on ourselves.

*Conformity to Norms.* The final major basis of self-presentation involves conformity to social rules and norms. Socially, we act in certain ways because we are taught that this is appropriate behavior, either in general intercourse or with reference to specific statuses. Many of these rules of self-presentation are implicit, operating below the level of conscious awareness, but this fact only accentuates their sociological importance.

A classic example of such social rules is the ritualistic deference expressed in traditional Oriental cultures. The Chinese and Japanese developed highly stylized and skilled techniques of self-abasement, characterized by the effort to present oneself as utterly worthless and the other as the most noble and worthy of mankind. In such interaction, of course, neither member believes that what he says about himself or the other is true; these are simply norms of courtesy.

The purposes of such norms vary. Some, such as the universal norm of *tact*, are intended essentially to protect the self-esteem of all members involved in the interaction. Others are designed to lubricate social relations in order to effect other ends. Still others are intended to reaffirm the extant status system. What all these norms have in common is the fact that they are fairly standardized ways of presenting oneself in interaction with others; such rules may be normative within the society as a whole or may be shared practices within some subgroup of it.

In addition to general behavioral prescriptions, there are specific social role demands imposed on the incumbent of a given status. Consider the doctor. The functional aspect of this role is to provide medical care for his patients and generally to act in the interests of their health. Beyond that, however, the doctor's role involves preferred but not absolutely obligatory behavior. Thus, the doctor should comport himself with dignity, not coarsely; should exude confidence, not insecurity; should show sympathy for suffering, not indifference; should devote himself day and night to the welfare of his patients; and so on. Although it is acceptable for the businessman's activities to be directed solely toward the acquisition of money, the self presented by the doctor is expected to be different.

That the self we attempt to set forth in social interaction does not necessarily correspond to the view we hold of ourselves does not mean that the presenting self and the extant self-concept are necessarily in conflict. No one can in large measure or in basic respects put on an act all the time; the strain would be intolerable. Hence, in general people will select presenting selves that are not only functional but also are comfortable and congenial, suited to their dispositions and consonant with their extant self-concepts. There is, then, a core of self-presentation which is trans-situational. But this is not to deny that the presenting self shows much greater variability than the extant self-concept or desired self-concept.

If, as suggested at the outset of this chapter, the self-concept is viewed as the totality of the individual's thoughts and feelings with reference to himself as an object, then it is apparent that the presenting self qualifies for inclusion. Implicitly or explicitly, the presenting self is a product of intention or decision; how we attempt to act is essentially under our control. Such a presenting self is predicated on the individual's ability to stand outside himself and to view himself as an object, to make decisions about that object and to carry out those decisions in speech and action. To some extent, every person is his own puppet master; our musculature submits to the authority of our minds.

In sum, we have suggested that the three major regions of the self-concept are the extant self, desired self, and presenting self. So conceived, the self-concept is largely a cognitive structure, a set of ideas about something. It exists on different planes and includes all these planes. But it is not simply an inert lump of knowledge, akin to known facts about the annual yield of corn in Iowa or the total square miles covered by the national forests. On the contrary, the self-concept matters, makes a difference, is motivated. People not only have self-concepts but they also have feelings and wishes about these self-concepts. It is therefore important to consider the thrust behind the self-concept—the self-concept motives—as well as certain principles underlying its formation.

## NOTES

1. Among other useful conceptual contributions are Bem (1965, 1967, 1972), Sherif (1968), Brissett (1972), Epstein (1973), Wells and Marwell (1976), Gergen (1971), and Gordon (1968, 1974, 1976).

2. This definition is similar to that of Rogers: "The self-structure is an organized configuration of perceptions of the self which are admissible to awareness" (1951: 501).

3. Best epitomized in the epigram: "There is more to the self than Mead's 'the I'."

4. In *Pudd'nhead Wilson*, Mark Twain remarked on the wide gap between logical and social definition. Speaking of the slave, Roxy, he observed: "From Roxy's manner of speech a stranger would have expected her to be black, but she was not. Only one-sixteenth of her was black, and that sixteenth did not show. . . . Her complexion was very fair. . . . To all intents and purposes Roxy was as white as anybody, but the one-sixteenth of her which was black outvoted the other fifteen parts and made her a negro. She was a slave and salable as such. Her child was thirty-one parts white, and he, too, was a slave. . . ."

5. The large volume of body-image research by Fisher and Cleveland (1958) is less concerned with the individual's picture of his own body than with how people's ways of thinking about the body may reflect underlying attitudes.

6. A well-known example is the favorability-unfavorability score of the Gough and Heilbrun (1965) Adjective Check List, based on 75 favorable and 75 unfavorable adjectives.

7. This view was advanced by Kuhn and McPartland (1954) in the first systematic, empirical self-concept study conducted by sociologists.

8. A recent book on "exceptionality" (Mordock, 1975) considers four major groupings of exceptional children. The first is concerned with incapacitating disorders—cerebral palsy, aphasia, mental retardation, schizophrenia, and autism. The second deals with "identifiable handicapping conditions"—hearing loss, visual impairment, speech handicaps, and chronic diseases. The third discusses "subtle handicapping conditions"—neurosis (hysteria, phobia), minimal brain dysfunction, learning disabilities. The fourth treats intellectual extremes, both the mildly retarded and the intellectually gifted. In other words, of 14 classificatory categories of children cited above, one refers to the intellectually advantaged, the remaining 13 to the disturbed, disordered, or disadvantaged. The term "exceptional" which, to most people's minds, connotes the first, is appropriate to cover the remainder.

9. It is precisely the *absence* of this feeling that Marx described as alienation, in its psychological sense. The work has no ego-involvement in the product of his own labor

because he neither owns nor controls the product, because he does not own the machinery or plant, and because the course of his efforts are directed by external, rather than internal, forces. To say that he is alienated from the product of his labor is to say that the sense of self is detached from the product.

10. This is not necessarily meant literally, as a physical photograph, but as a concept including selected physical characteristics, social identity elements, and dispositions, along with an associated "style of life."

11. The outstanding description of the social rules of self-presentation appears in the work of Erving Goffman (1955; 1956). His signal contribution has been to point out that the individual has a certain image of the type of person he should appear to be in a given social situation; that this image frequently corresponds neither to what the individual actually is, thinks of himself as, or would like to think of himself as; that it is defined as socially appropriate in certain cultures either in general or with reference to the social statuses of the participants; and that these norms are often below the level of explicit awareness, representing shared understandings which guide behavior but which, remaining unverbalized, are largely unconscious. Certain of Goffman's followers, emphasizing the critical significance of language in social life, have stressed the implicit language rules involved in the protection and enhancement of the self.

# 2

---

# The Self-Concept:
# Motives and Principles

---

IF the self-concept were solely the complex structure described in the previous chapter, it would still exercise an immense influence on our lives. But it is more than that: it is also a motivational system. Certain aspects, components, or dimensions are *desired*, serving as spurs for action as well as guides for perception.

## Self-Concept Motives

Discussions of self-concept motives generally agree that "self-seeking and self-preservation" (James, 1890) or the "maintenance or enhancement of the self" (Snygg and Combs, 1949) are central to the individual's motivational system. Although commonly joined, it should be pointed out that two separate motives are involved. The first is the self-esteem motive—the wish to think well of oneself. The second is the self-consistency motive—the wish to protect the self-concept against change or to maintain one's self-picture. Schwartz and Stryker (1971: 2) hold that "(1) persons seek to create and maintain stable, coherent identities; (2) persons prefer to evaluate their identities positively." Although these two motives rarely conflict, there are occasions, as we shall see, where they may work at cross-purposes. Both motives, we believe, have powerful emotional and behavioral consequences; furthermore, they are not simply passively preferred but also actively sought.

### SELF-ESTEEM

Implicitly or explicitly, there is widespread agreement with Howard Kaplan's contention that "the self-esteem motive is universally and characteristically . . . a dominant motive in the individual's motivational system" (1975: 10). As Gordon Allport (1961: 155–6) observes: "If we are to hold to the theory of multiple drives at all, we must at least admit that the ego drive (or pride or desire for approval—call it what you will) takes precedence over all other drives."

In the present discussion, self-esteem signifies a positive or negative orientation toward an object. When we characterize a person as having high self-esteem, we are *not* referring to feelings of superiority, in the sense of arrogance, conceit, contempt for others, overweening pride; we mean, rather, that he has self-respect, considers himself a person of worth. Appreciating his own merits, he nonetheless recognizes his faults, faults that he hopes and expects to overcome. The person with high self-esteem has *philotimo*, not *hubris*; he does not necessarily consider himself better than most others but neither does he consider himself worse. The term "low self-esteem" does not suffer from this dual connotation. It means that the individual lacks respect for himself, considers himself unworthy, inadequate, or otherwise seriously deficient as a person.

Certain depth psychologists have gone so far as to contend that self-esteem problems are at the heart of the neurotic process. Angyal (1941) states: "In the neurotic development there are always a number of unfortunate circumstances that instill in the child a self-derogatory feeling. This involves on the one hand a feeling of weakness which discourages him from the free expression of his wish for mastery, and on the other a feeling that there is something wrong with him and that, therefore, he cannot be loved. The whole complicated structure of neurosis appears to be founded on this secret feeling of worthlessness, that is, on the belief that one is inadequate to master the situations that confront him and that he is undeserving of love."

Systematic quantitative data are entirely consistent with the psychoanalyst's clinical insights. One study (Rosenberg, 1965), based on a sample of over 5,000 high school juniors and seniors, showed that only 4 percent of those with the highest self-esteem but fully 80 percent of those with the lowest self-esteem were highly depressed, according to a scale of "depressive affect" ($r = .3008$). Furthermore, only 19 percent of those with the highest self-esteem showed a relatively large number of psychophysiological indicators of anxiety (hand trembling; heart pounding; pressures or pains in the head; hands sweating; dizziness; etc.) compared to 69 percent of those with the lowest self-esteem ($r = .4848$). Similarly, Kaplan and Pokorny (1969) showed "self-derogation" to be related to physical indicators of anxiety, to depressive affect, and to the use of psychiatric assistance.

The data from Bachman's (1970) study of 2,213 tenth-grade boys throughout the country are equally persuasive. In this sample, he found the following correlations between self-esteem and a number of measures of emotional disturbance: negative affective states ($-.52$); "happiness" ($+.54$); somatic symptoms ($-.34$); and impulse to aggression [1] ($-.34$) (1970: 122). Persons with high self-esteem are also decidedly more likely to express high satisfaction with life (Crandall, 1973). The well-known work of Beck (1967) indicates that low self-esteem is one of the distinguishing features of depression. In addition, Luck and Heiss (1972) found their measure of global self-esteem to be significantly related to submissiveness, depression, psychic anxiety, somatic anxiety, autonomic anxiety, maladjustment, and vulnerability among adult white males. Jahoda (1958) holds that a high level of self-acceptance or self-respect is an important component of "positive mental health." The fact that probably more research has been devoted to self-esteem than to all the other aspects of the self-concept combined (Wylie, 1961, 1974) is no doubt attributable to the great relevance of self-esteem for emotional disturbance (Rogers, 1951; Turner and Vanderlippe, 1958; Wylie, 1961, Chap. IV).[2]

One of the outstanding products of Freud's genius was his discovery of the psychoanalytic defense mechanisms. It has been convincingly argued (Murphy, 1947; Allport, 1955; Hilgard, 1949), however, that these defense mechanisms are employed largely in the service of self-esteem protection and enhancement. Examples come easily to mind. *Rationalization* [3] involves finding a socially acceptable or admirable explanation of our behavior that might otherwise be condemned. *Compensation* represents an effort to overcome the damage inflicted on self-esteem as a consequence of failure in one area by extraordinary achievement in that or another area. *Projection* involves attributing to others certain undesirable characteristics or wishes which in fact characterize the self, but which, if recognized, would be offensive to self-esteem. A well-known manifestation of *displacement* is scapegoating, used by people who, frustrated and humiliated by those more powerful, seek to boost their own self-esteem by asserting their superiority over others. *Reaction formation* involves emphasizing feelings or characteristics which are precisely the reverse of certain undesirable characteristics of the actual self. (For example, the mother who unconsciously hates her child is effusively loving; the man with unconscious homosexual tendencies becomes a Don Juan.) *Repression* involves thrusting into the unconscious libidinal or aggressive impulses which, if recognized, would offend self-esteem (for example, wish to destroy father, copulate with mother). To a substantial extent, these mechanisms have as their objective the protection of self-esteem.

Only a motive of enormous power could explain the wide range of devices (of which the Freudian mechanisms are only a sample) marshalled by individuals of every intellectual caliber in defense of self-esteem. As Brendan Gill (1975: 4), after being shown up publicly in a mistake, expressed it:

Nevertheless, I am always so ready to take a favorable view of my powers that even when I am caught out and made a fool of, I manage to twist this circumstance about until it becomes a proof of how exceptional I am. The ingenuity we practice in order to appear admirable to ourselves would suffice to invent the telephone twice over on a rainy summer morning.

Furthermore, the individual does not simply ensconce himself behind his lines of defense but he ventures forth actively and aggressively. He does not merely protect his reputation, he also searches for fame; he does not merely strive to avoid others' negative opinions but works equally to elicit their positive opinions. This is part, though not all, of what Allport (1955) means by "propriate striving" and James means by the self as a "fighter for ends."

The self-esteem motive is thus one of the most powerful in the human repertoire. Curiously, there is no agreement on why this should be so. Some writers, such as Kaplan (1975), hold that the wish for positive self-attitudes is associated with certain pleasurable and rewarding experiences of childhood. Gergen (1971) concurs, viewing the desire for high self-esteem as the outcome of the process of secondary reinforcement. Aspects, qualities, or characteristics of the self which have proved useful in producing pleasurable or satisfying outcomes themselves come to be valued.

But other writers, such as James, consider the self-esteem motive to be more fundamental—"direct and elementary endowments of our nature. . . . the emotions . . . of self-satisfaction and abasement are of a unique sort, each as worthy to be classed as a primitive emotional species as are, for example, rage or pain" (1890: 307). Similarly, in advancing their phenomenological view of the self, Snygg and Combs (1949) postulate that the protection and enhancement of the self are themselves prime motives, not reducible to more elementary drives. McDougall's (1932) theory of sentiments postulates "self-regard" as the master sentiment, the sentiment to which all others are subordinated. In these views, the self-esteem motive rests on its own foundations; high self-esteem is innately satisfying and pleasurable, low self-esteem the opposite. A major determinant of human thought and behavior and a prime motive in human striving, then, is the drive to protect and enhance one's self-esteem.

## SELF-CONSISTENCY

Side by side with the self-esteem motive stands what is sometimes called the "self-consistency" motive. Various terms appear in the literature expressing this idea: protection of the self, self-preservation, maintenance of the self, and self-concept stability. Although Lecky's (1945) term "self-consistency" is not entirely a felicitous one, at times referring to the motive to maintain a stable self-concept, at other times to the logical articulation of the elements, we shall follow the current general practice of using the term in its former sense.

Self-consistency refers to the motive to act in accordance with the self-concept and to maintain it intact in the face of potentially challenging evidence. People behave in a fashion consistent with the pictures they hold of themselves and interpret any experience contradictory to this self-picture as a threat. In Lecky's words: "The individual's conception of himself is the central axiom of his whole life theory" (1945: 264–5). ". . . All of an individual's values are organized into a single system the preservation of whose integrity is essential. The nucleus of the system, around which the rest of the system revolves, is the individual's valuation of himself. The individual sees the world from his own viewpoint, with himself as the center. Any value entering the system which is inconsistent with the individual's valuation of himself cannot be assimilated; it meets with resistance and is, unless a general reorganization occurs, to be rejected" (1945: 152–3).

Although it is reasonable to expect instability of the self-concept to be psychologically distressing, the evidence in this regard is very limited. In a study of adolescents, we found that the stability of the self-concept was strongly related to certain psychophysiological indicators of anxiety (Rosenberg, 1965: 143). Since self-concept stability is related to self-esteem, however, one cannot be certain that the associated factor of self-esteem is not responsible for the observed relationship. We have examined this question in Bachman's nationwide study of tenth-grade boys. The correlation between self-concept stability and self-esteem was .2406. It is interesting, however, that the correlation of stability to a measure of "somatic symptoms" (similar to the above psychophysiological indicators) was .1875, and the correlation of stability to a score of "negative affective states" was .3719, *even when self-esteem was controlled* through partial correlation. Instability thus appears to be associated with signs of psychological disturbance independent of self-esteem.

The power and persistence of the self-consistency motive may be quite remarkable. People who have developed self-pictures early in life frequently continue to hold to these self-views long after the actual self has changed radically. Reports are common of people who, in childhood, were either very thin or very fat but have, in the course of years, either shed or accumulated pounds. Yet the slim person continues to think of himself as fat (or as a fat person who has lost weight), whereas the double-chinned still has an image of the self as a twig. Similarly, the person who grows gruff and irritable with the passing years may still think of himself as "basically" kindly, cheerful, and well-disposed; the behavior which has become chronic is either unrecognized or is perceived as a temporary aberration from the true self.

The reasons for expecting the self-consistency motive to be important are compelling. If we view the self-concept as an attitude toward an object, then one reason may be found in Allport's (1954: 44) discussion of the function of attitudes. "Without guiding attitudes the individual is confused and baffled. Some kind of preparation is essential before he can make a satisfactory observation, pass suitable judgment, or make any but the most primitive reflex type of response. Attitudes determine for each individual what he will see and hear, what he will think and what he will do. To borrow a phrase from William James, they 'engender meaning upon the world'; they are our methods of finding our way about in an ambiguous universe."

The general need to maintain stable attitudes is amplified enormously with respect to self-attitudes, for without some picture of what he is like, the individual is virtually immobilized. Insofar as he is an actor in any situation, he must operate on at least some implicit assumption of what kind of person he is and how others see him. If he considers himself weak, he will not undertake to lift a heavy object; if musically untalented, he will avoid musical training; if unattractive, he will not ask for a date; if unintelligent, he will not apply for graduate education; and so on. The important point is that the individual's decisions are based not on what he actually is but on what he thinks he is. These assumptions may be true or false, but they are decisive.

The self-concept is thus the individual's fundamental frame of reference, the foundation on which almost all his actions are predicated. It is small wonder that he is so eager to define his self-concept and, having reached a conclusion, struggles so ardently to defend and protect it against change. What psychoanalysts interpret as pathological resistance Lecky interprets as a healthy effort to maintain one's integrity, to be true to one's self-picture.

Self-esteem and self-consistency—enhancing and maintaining the self-concept—are thus two prime but distinct motives guiding human behavior. These twin motives ordinarily enjoy harmonious relations with one another; self-seeking and self-preservation are usually served by the same actions. But what if these motives clash; which emerges victorious? According to Lecky, the motive of consistency may override the self-enhancement drive. Taking the case of an intelligent student who is a poor speller, Lecky argues that in almost every case further tutoring fails, despite the student's ability. The reason is that in the past the individual has incorporated into his self-concept the idea that he is an incompetent speller and resists any evidence that would force him to alter that view.

That this may occur in particular cases, however, does not indicate whether the pattern is a general one. Recently, investigators have attempted to bring systematic evidence to bear on the question: which motive—self-esteem or self-consistency—is the more powerful?

Stephen Jones (1973) has summarized 16 studies which, in one way or another, examined this issue. According to Jones, self-esteem theory implies that we will like those people who think well of us and dislike those whose opinion of us is negative. Self-consistency theory holds that we will like those who see us as we see ourselves and dislike those whose view of us is different from our own.

The matter is complicated because these competing motives cannot be tested among those with high self-esteem. If our self-esteem is high, then our liking for someone who thinks well of us, and our disliking of someone who thinks ill of us, may be due to our wish to maintain *either* self-esteem *or* self-consistency. It is only if we have *low* self-esteem that the two motives yield different predictions. Self-esteem theory would hold that we like those who think well of us and dislike those who think ill of us, whereas self-consistency theory predicts that we like those who think ill of us (in agreement with our own view) and dislike those who think well of us (in contradiction to our own view).

Which theory do the data support? According to Jones (1973: 192): "... The evidence, in general, tends to favor self-esteem theory over self-consistency theory. Of the 16 investigations reviewed, 10 support self-esteem theory and ... there are serious problems of interpretation or replication with the experimental studies often cited as support for self-consistency theory."

These data, of course, do not settle the issue, as Jones himself observes. The individual with low self-esteem may agree that the other person is indeed right to hold him in low regard, but it is rather extreme to expect him to *like* the other person *because* the other denigrates him, or to *hate* the other person *because* he thinks well of him. People who tell us derogatory things about ourselves, even if we agree they are true, are rarely thanked for their frankness.

A study by Fitch sheds further light on the matter. According to him (1970: 311): "Two partially contradictory hypotheses may be derived from self theory. The first is that persons are motivated to perceive events in a way which *enhances* chronic self-esteem. The second is that persons are motivated to perceive events in a way which is *consistent* with chronic self-esteem." Subjects were first classified as having high or low self-esteem, were given a "test," were randomly informed that they had succeeded or failed, and were then asked what accounted for their performance. Following attribution theory, these explanations could be attributed either to internal causes (their own ability and effort) or to external causes (idio-syncracies of the test, or physical or mental conditions). Those with high self-esteem who had "succeeded" explained their success in terms of personal merit, whereas those who had "failed" attributed this performance to external or accidental factors.[4] Both interpreta-tions protected self-esteem and maintained self-consistency. But what about those who initially started with low self-esteem? In this group, both those who "succeeded" and those who "failed" were about equally likely to attribute their performance to internal factors. In other words, the low self-esteem subject who "fails" is apparently more likely than the high self-esteem subject to believe that he de-serves to fail. Overall, the self-esteem and self-consistency motives appear to have about equal strength in this study.

Perhaps the most interesting finding is that people with low self-esteem, or those holding negative attitudes toward specific self-con-cept components, may nevertheless adamantly refuse to accept in-formation that will improve their self-esteem. One reason, suggested by Epstein (1973), is that people may *retain* low self-esteem in or-der to *protect* self-esteem. This is the well-known "failure of nerve." In a sense, a person who expects nothing from himself cannot fail, since his weak performance meets his meager expectations. In Thomas Carlyle's words, "Make thy claim of wages a zero, then hast thou the world under thy feet" (quoted in James, 1890). The low self-esteem person may be reluctant to believe that he is more intelligent, attractive, or masterful than he currently assumes be-cause, acting on these assumptions, he might find his hopes dashed, his aspirations frustrated. As we shall indicate in chapter 11, re-search consistently shows that people establish their aspiration levels in a fashion designed to maximize self-esteem. Thus, the low self-esteem person who maintains self-consistency by setting low aspira-tions and expecting to perform poorly is at the same time protecting his self-esteem by avoiding failure.

Paradoxically, then, the incorporation of negative components into the self-concept may actually enhance self-esteem. For exam-ple, if someone is assigned a negative label, it may serve the interests of self-esteem to accept rather than to reject it. The Hammersmith and Weinberg (1973) study of male homosexuals demonstrates this point. The subjects were asked whether they accepted or rejected the homosexual condition, that is, whether they wished they were not homosexual and whether they sought to overcome their homosexuality. The data clearly showed that those respondents who accepted their homosexuality had higher self-esteem (in fact, self-esteem equal to that of heterosexuals) than those who rejected it. Although we cannot generalize about those conditions under which the principle holds, it is plain that at least in some situations the acceptance of the negative label may protect and enhance self-esteem.

But self-esteem aside, people are motivated to hold to their self-pictures, for they are lost without them. The question of whether self-esteem or self-consistency is the more powerful motive may thus not be a very meaningful one, since this may depend on whether we are speaking of the self as a whole or in terms of its specific components, whether the particular component is central or peripheral to the self-concept, and so on. But it is relevant to note that when writers speak of the maintenance and enhancement of the self (or self-seeking and self-preservation), they are speaking of two extremely powerful motives, both playing major roles in human thought, feeling, and behavior.

## Self-Concept Formation: Four Principles

Throughout this work, four principles will be advanced in an effort to explain the diverse phenomena to be considered. These principles, we believe, underlie most of the theoretical reasoning employed in the literature to understand the bearing of interpersonal and social structural processes on the self-concept. Not all principles, of course, have equal relevance to all phenomena; in some cases, certain principles are suitably invoked to cover given empirical facts, while in other cases, different principles apply. Although these principles may lead to faulty empirical predictions, either because they are misapplied or are applied without refinement, we suggest that the following four principles bring an impressive level of coherence to a diversity of empirical data: reflected appraisals, social comparisons, self-attribution, and psychological centrality.

### REFLECTED APPRAISALS

According to Harry Stack Sullivan (1947: 10): "The self may be said to be made up of reflected appraisals. If these were chiefly

derogatory, as in the case of an unwanted child who was never loved
. . . then the self-dynamism will itself be chiefly derogatory." Although we shall see that the self-concept is made up of more than
reflected appraisals, their significance for the self-concept can scarcely
be overestimated. Reduced to essentials, this principle holds that
people, as social animals, are deeply influenced by the attitudes of
others toward the self and that, in the course of time, they come to
view themselves as they are viewed by others. This principle is fundamental to any understanding of the relationship of social structure and social interaction to the self-concept. Furthermore, it is
empirically true and directly relevant to our present concerns.

Straightforward, and almost trivial, as the idea appears, it sometimes combines related ideas which are more usefully distinguished.
Specifically, these are the principles of (1) direct reflections, (2)
perceived selves, and (3) the generalized other. The first refers to
how particular others view us, the second to how we believe they
view us, and the third to the attitudes of the community as a whole;
these are internalized in the "me" and serve as a perspective for
viewing the self.

The principle of direct reflections, holding that the self-concept is
largely shaped by the responses of others, was set forth felicitously
by Thorstein Veblen (1934: 30):

Those members of the community who fall short of this, somewhat indefinite, normal degree of prowess or of property suffer in the esteem of
their fellow man; and consequently they suffer also in their own esteem,
since the usual basis of self-respect is the respect by one's neighbors. Only
individuals with an aberrant temperament can in the long run retain their
self-esteem in the face of the disesteem of their fellows.

In general, a reasonably good level of correspondence between
others' views of us and our own is completely indispensable for adjustment to society. In situations where a gross and fundamental
discrepancy exists, the person is considered simply psychotic. There
is nothing wrong with believing one is Napoleon or Alexander the
Great if the rest of society believes equally that one is. But, even in
less extreme cases, any substantial discordance between our selfview and the view others hold of us may generate considerable
difficulty. Our claims for deference, based on our assumed exalted
faculties, will scarcely be honored by others if they do not share our
own high regard for ourselves. Our behavior, predicated on our irresistible charm, will be deemed ludicrous by those who find our charm
all too easy to resist. The difficulties that arise as a consequence of

discordant definitions of the self are familiar. For example, when the adolescent sees himself as a mature, responsible young adult while the parent continues to see him as an irresponsible child ("eat your carrots, they're good for you"), tempers are sure to flare.

Because it is so essential to know what we are like if we are to have any firm basis for action, and because it is so difficult to arrive at this knowledge, other people's judgments of us matter enormously; indeed, there is probably no more critical and significant source of information about ourselves than other people's views of us. We need *consensual validation* of our self-concepts. But the matter is still more fundamental, for the very sense of self arises through the process of adopting the attitudes of others toward the self.

The most subtle and sophisticated exposition of this viewpoint is to be found in the work of George Herbert Mead (1934). Mead pointed out that the fundamental social process of communication requires the individual to "take the role of the other." His "self" emerges as he comes to respond to himself from the standpoint of others. "The individual experiences himself as such not directly, but only indirectly, from the particular standpoints of other individual members of the same social group, or from the generalized standpoint of the social group as a whole to which he belongs" (Mead, 1934: 138).

If the process of communication obliges the individual to "become an object of himself . . . by taking the attitudes of other individuals toward himself," it is reasonable to think that others' *evaluations* will affect the individual's *self-evaluation*. Mead's (1934: 68) conclusion that "We are more or less unconsciously seeing ourselves as others see us" should suggest a general correspondence between others' attitudes toward us and our attitudes toward ourselves.

Lest there be any misunderstanding on this point, the principle of direct reflections suggested here is an *inference* from Mead's theory; to say that we see ourselves from others' viewpoints is not the same as saying that our self-views correspond precisely to the views others hold of us. Nevertheless, the principle appears to us to be implicit in his work and, as a matter of fact, has received strong and consistent empirical confirmation over nearly a quarter of a century (Miyamoto and Dornbusch, 1956; Reeder, Donohue, and Biblarz, 1960; Backman and Secord, 1962; Backman, Secord and Pierce, 1963; Brookover, Thomas, and Paterson, 1964; Videbeck, 1960; Manis,

1955; Sherwood, 1965, 1967; Deutsch and Solomon, 1959; Quarantelli and Cooper, 1966; and many more). Because this literature is large, a brief description of the first empirical test of Mead's hypothesis—that of Miyamoto and Dornbusch—may serve as representative of the remainder.

In their 1956 study, Miyamoto and Dornbusch collected data from 195 subjects divided into 10 groups ranging in size from 8 to 48 persons. Four measures were used: (1) Self-conception. Each subject rated himself on a 5-point scale with regard to four characteristics—intelligence, self-confidence, physical attractiveness, and likeableness. (2) Actual responses of others. On the same 5-point scale, each person rated every other member of his own group in terms of these 4 characteristics. (3) Perceived responses of others. Each person predicted how every other member of his group would rate him on these four scales. (4) Generalized other. Each subject was asked how he perceived most persons as viewing him in terms of these four characteristics.

The specific test of the principle of direct reflection was whether those who rated themselves favorably were more likely to be rated favorably by others. This question was examined in each of the 10 groups for each of the 4 characteristics. Of the 40 possible tests, it turned out, the hypothesis was supported in 35 cases, not supported in 4 cases, and tied in one. These data clearly show that the individual tends to see himself as he is actually seen by others. Subsequent research in the succeeding two decades has consistently supported this finding.[5]

The second idea frequently subsumed under the reflected appraisals principle is that of the "perceived self." Although Cooley's (1912) famous term "the looking-glass self" is frequently interpreted to refer to direct reflections, Cooley himself stressed that this term was not entirely apt. The "looking-glass self," he held, involves "the *imagination* of our appearance to the other person and the *imagination* of his judgment of that appearance," as well as some self-feeling, such as pride or mortification. Thus "the thing that moves us to pride or shame is not the mere mechanical reflection of ourselves, but an imputed sentiment, the imagined effect of this reflection upon another mind" (Cooley, 1912: 152). It is thus not others' attitudes toward us but our *perception* of their attitudes that is critical for self-concept formation.

Empirical data strongly and unequivocally support this view. The evidence clearly demonstrates that the relationship between the self-concept and the "perceived self" is a strong one—in fact, considerably stronger than the relationship between the self-concept and the "social self" [6] (what others actually do think of us) (Miyamoto and Dornbusch, 1956; Reeder, Donohue, and Biblarz, 1960; Sherwood, 1965). As one example, Reeder, Donohue, and Biblarz (1960) conducted a study of 54 military personnel who had been divided into 9 groups of from 5–9 members each. Subjects were asked to rate themselves; to rank every member of the group; and to indicate how he thought every other group member would rate him in terms of leadership. (The same questions were asked with regard to rank as a "good worker" in the group.) The results show a very close correspondence between how the individual believed other soldiers rated him as a leader and worker and his own self-rating. Of the 54 soldiers, fully 46 believed the group rated them as they rated themselves; only 8 anticipated discrepancies. With regard to "worker" characteristics, the correspondence was lower (38 out of 54) but still substantial. The point is that both of these relationships were appreciably stronger than the relationship between the self-concept and the social self (what others actually thought of the individual).

Similarly, in the Miyamoto and Dornbusch (1956) study cited above, the investigators examined the relationship of the *perceived* responses of others and the individual's self-evaluation. In this case, the prediction was that "the mean of the perceived responses of others will be higher for those persons with a high self-rating than for those with a low self-rating." Testing this hypothesis for 4 characteristics in 10 groups, the hypothesis was supported 40 out of 40 times. Again, the relationship between the perceived self and the self-concept was stronger than the relationship between the social self and the self-concept.

The third sense in which the attitude of others is said to affect the self-concept is that of Mead's "me," based largely on the generalized other. The self, Mead stressed, arises out of social experience, particularly social interaction. The process of communication obliges the individual to adopt the attitude of the other toward the self and to see himself from their standpoint or perspective. In the well-known example of the baseball game, Mead points out that the individual cannot play the role of third baseman without having internalized the attitudes of all the others engaged in this interaction—the catcher, the pitcher, the second baseman—toward third basemen. Hence, he must incorporate into himself the attitudes of all the others participating in this organized social interaction if he is to play his own role effectively. Indeed, the very universality of thought is "the result of the given individual taking the attitudes of others toward himself, and of his finally crystallizing all these particular attitudes into a single attitude or standpoint which may be called that of the 'generalized other'" (1934: 90). The individual condemning himself for an immoral act, for example (usually based on the action of the "I"—the spontaneous and unpredictable aspect of the self), does so as a consequence of having internalized in the "me" the universal attitudes of condemnation toward certain behavior. The individual's self-concept is shaped here by the attitudes of others, not as a direct reflection of these attitudes, but by applying to the self the attitudes of the society as a whole.

All three concepts—direct reflections, perceived self, and the generalized other—are concerned with the role of other people's attitudes in shaping self-concepts. Though conceptually distinct, each of these processes may, in its own way, produce the same result. Take the example of someone who has cheated on an examination. If those who know he has done so treat him with contempt, he might experience low self-esteem as a result of direct reflections. If, on the other hand, he *infers* that they feel contempt for him (perhaps they appear to avoid him or refuse to meet his eye) when in fact they have no knowledge of his peccadilloes, his resulting low self-esteem would be the product of the perceived self.[7] Finally, if he condemns himself (with consequent self-esteem reduction) because he has internalized the value system of particular others (perhaps his mother or father) or of the society as a whole, then, even though others are not directly involved, his self-attitudes would still be governed by other people's perspectives or standpoints. All three processes are expressions of the principle of reflected appraisals and testify to the importance of others' attitudes toward us in determining our self-concepts.

SOCIAL COMPARISONS

The principle of social comparison is fundamental to self-concept formation and is a major component of what has come to be known as social evaluation theory. As described by Pettigrew (1967: 243): "The basic tenet of social evaluation theory is that human beings learn about themselves by comparing themselves to others. A second tenet is that the process of social evaluation leads to positive, neutral, or negative self-ratings which are relative to the standards set by the individuals employed for comparison."

Our present interest centers on the fact that people judge and evaluate themselves by comparing themselves to certain individuals, groups, or social categories. This is not to suggest that they may not *also* compare themselves with other standards. Thibaut and Kelley (1959) note that people may evaluate themselves in light of their own past performance; James (1890) holds that the mature individual may compare himself with the standards set by the internalized "ideal judge;" and obviously people may compare themselves with many other standards such as the idealized image, the committed image, and the moral image. In the present discussion, however, we restrict our focus to the individual's comparison of himself with what Pettigrew (1967) calls "referent individuals" and "reference groups."

Two types of social comparison may usefully be distinguished. One marks individuals as superior or inferior to one another in terms of some criterion of excellence, merit, or virtue. Smarter or duller, weaker or stronger, handsomer or homelier are comparative terms requiring a relative judgment both of others and of the self. Without undergoing the slightest physical transformation, we are metamorphosed from the weakest to the strongest person simply by shifting our basis for comparison.

The other type of social comparison is normative, and refers primarily to deviance or conformity. Here the issue is not whether one is *better* or *worse* but whether one is the *same* or *different*.[8] For example, the adolescent excoriated in the home for nonconformity to certain rules or values is applauded by his peers for the identical behavior. Conformity or deviance do not inhere in the behavior as such but in its relation to the norms of the particular environment. As we shall indicate in chapter 4, both types of social comparison— superiority-inferiority and conformity-deviance—have significant self-esteem consequences.

The social comparison principle is one of unquestionable power, making meaningful psychological sense of a range of disparate phenomena. But its great strength, which lies in its generality, is also its great weakness, which is its excessive flexibility and lack of specificity. One reason that the social comparison principle "works" so well is that it is always easy for the *investigator* to think of some plausible group with which the individual is presumably comparing himself, and to explain his findings in these terms. As noted by Merton and Kitt (1950: 49):

Since both membership groups and non-membership groups, in-groups and out-groups, have in fact been taken as assumed social frames of reference . . . this at once leads to a general question of central importance to a developing theory of reference group behavior: under which conditions are associates within one's own groups taken as a frame of reference for self-evaluation . . . and under which conditions do out-groups or non-membership groups provide the significant frame of reference?

Additional questions are: with reference to *what* characteristics is the individual comparing himself to others, and how *important* are these attributes in the individual's system of self-values? Although there is much still to be learned in this area, social comparison theory remains one of the most exciting topics in social psychology, and is of great relevance to the present work. The ability of social comparison theory to make sense of data which, in the absence of the theory, would be particularly baffling, can be illustrated by several examples. The first is what James Davis (1966) has called the "frog-pond effect." Davis's investigation is based on an examination of the relationship of college performance to career choice among 35,000 college graduates. Colleges, of course, vary widely in average academic level so that a person who does well in a school with low standards is unlikely to be the intellectual equal of one who does equally well in a highly select school (as determined by National Merit Scholarship tests). Despite the considerable difference in intellectual quality, however, Davis found that aspirations to enter such fields as medicine, law, science, humanities, etc. were surprisingly similar across schools. The reason, it turns out, is that the college students, in making their career choices, are comparing themselves chiefly with those *in their own* schools, not with those throughout the country with whom they will actually compete.

In a study of school pupils in Baltimore City, to be described in the next chapter, it was found that the higher the pupil's school marks, the higher his global self-esteem tended to be.[9] This principle held for both black and white children. In addition, the data showed that in this study black secondary school pupils attending predominantly white schools had somewhat higher grade averages than those in predominantly black schools. But though high marks raised self-esteem, and black children in white schools had higher marks than those in black schools, their self-esteem, instead of being higher, was lower. The reason was that the desegregated black children were comparing their school performance not with that of the segregated black children but rather with that of the white children in their own schools where, it turned out, they suffered by comparison.

If we focus on standardized achievement tests instead of school marks and on "academic self-concept" instead of global self-esteem, the same results appear. The well-known Coleman Report (Coleman *et al.*, 1966), a nationwide survey of over 600,000 school pupils, found that "as the proportion [of] white [children to black] in the school increases, the black child's academic self-concept *decreases*" despite the fact that the objective achievement scores of the black child in a predominantly white school are *superior* to those of the black child in the black school.

Since, to an important extent, people judge themselves by comparing themselves to others, the principle of social comparison has special relevance for the self-concept. But who these referent others are, and with regard to what characteristics individuals compare themselves to others—these will be detailed in succeeding chapters. A number of unexpected findings regarding the relationship between social structure and the self-concept will appear more reasonable when viewed from the social comparison perspective.

### SELF-ATTRIBUTION

"Self-perception theory" was originally advanced by Bem (1967) as an alternative explanation for the findings of cognitive dissonance studies. It is founded essentially on the "radical behaviorism" of B. F. Skinner. This theory advances the rather imaginative idea that even reports of inner states (such as hunger, anger, excitement, tension, sympathy), which are ordinarily assumed to be based on purely private internal stimuli, may in reality reflect past training in the application of descriptive terms to certain overt behavior—whether of other people or of the self. For example people are taught to apply the term "anger" to certain types of emotional behavior, the term "sympathy" to other types of behavior, and so on. Bem's argument is that the individual's self-descriptions of his emotional states may thus essentially be based upon his observation of his own overt behavior and the conditions under which it occurs.

The bases for applying self-descriptive statements to one's own internal states are, according to this theory, not essentially different from the bases used by other people to describe our internal states.

Bem gives the example of an individual who, asked whether he likes brown bread, replies "I guess I do, I'm always eating it." This individual is basing his statement on the same information as his wife when she, asked the same question about him, replies, "I guess he does, he's always eating it." Similarly, the individual who, after devouring three sandwiches and two pieces of pie, comments "I guess I was hungrier than I thought" is describing his inner state on the basis of his observation of his own behavior rather than on physiological experiences.

The more widely recognized "attribution theory" has tended to subsume Bem's self-perception theory as a special case. According to Kelley (1967: 193): "Attribution refers to the process of inferring or perceiving the dispositional properties of entities in the environment." Founded on Heider's (1958) views on phenomenal causality, it is primarily concerned with how the naive individual thinks in terms of causes. Specifically, how do people ordinarily explain what they observe? Some explanations may be external (the cause lies in some aspect of the environment), while others are internal (the cause is to be found in motives, intentions, or dispositions of the actor). The central point is that one of the "entities in the environment" to which attributions are made is the self. If he were to focus on self-attribution, the attribution theorist would be interested in understanding the *bases* on which people draw conclusions about *their own* motives or underlying characteristics and how they go about verifying their tentative conclusions.[10] The attribution theorist would agree with the self-perception theorist that the individual's observation of his overt behavior represents a major basis for drawing conclusions about his inner motives, states, or traits, although it is unlikely that he would insist as rigidly as the behaviorist on excluding internal stimuli as sources of information.

In focusing on such subtle and elusive phenomena as physiological states, motives, wishes, and intentions, it is easy to overlook the more obvious and indisputable cases of self-attribution. The individual certainly does draw conclusions about his dispositions—especially, but not exclusively, abilities or other types of competence—in part on the basis of observing his own behavior or its outcomes. An example is the child who consistently does well on spelling tests and consequently concludes that he is a good speller; this conclusion is reached not primarily by consulting his inner experience but by observing his behavior or its outcomes. The person who tries skiing for the first time and finds his efforts crowned with success changes his view of his skiing ability. Similarly, the individual whose every effort to do house repairs results in disaster concludes that he lacks mechanical aptitude. The person who achieves success in business, the theater, or the academy concludes, by reflecting on his achievements, that he possesses certain underlying talents.

This theory is certainly compelling and important. Nor is it necessarily inconsistent with any other theories. It is the "I"—the spontaneous, unpredictable part of the self—which, at the moment of action in the baseball game, leaps and spears the sharply hit line drive, and the "me" which, internalizing the general views toward such behavior, judges the catch to be brilliant. Whatever the ultimate theoretical resolution of the question of how we draw conclusions about certain internal states, there can be little doubt that we draw conclusions about ourselves largely by observing our behavior and its outcomes.

Perhaps the most consistent empirical support for the self-attribution principle is the finding that youngsters who do well in school are more likely to hold high "academic self-concepts" (think they are good students or that they are smart) (Purkey, 1970). In a study of New York State high school juniors and seniors (to be described later), the relation (contingency coefficient) between school marks and regarding oneself as a "good student in school" was .52. Bachman's (1970: 242) nationwide study of tenth grade boys found the correlation of self-concept of school ability to school marks to be $r = .4817$. Brookover et al. (1964) reported an association of $r = .57$ between school marks and academic self-concept for both males and females. How good a student the child thinks he is obviously depends largely on how well he has done in school.

In our view, the radical behaviorist position of self-perception theory, holding that people are basically applying descriptive statements, learned from earlier experience, to observed behavior and its associated stimulus conditions is too strictly divorced from phenomenal reality to advance self-concept understanding. To deny that we feel, and to know that we feel (either directly or upon reflection) anger, jealousy, euphoria, or excruciating boredom even as we maintain a bland and unexpressive countenance is to substitute theoretical purity for human experience. But it is undeniably the case that people do draw conclusions about how smart, kind, generous, or musically talented they are in considerable part on the basis of observing their own actions and its outcomes. It is this process we shall have in mind in speaking of "self-attribution."

### PSYCHOLOGICAL CENTRALITY

As noted in the earlier discussion of psychological centrality, this principle holds that the self-concept is not a *collection* but an *organization* of parts, pieces, and components and that these are hierarchically organized and interrelated in complex ways. Not only are certain dispositions—intelligence, morality, honesty, courtesy—differentially central to our concerns, but so are certain social identity elements (such as black, Protestant, father, machinist) and ego-extensions. Unfortunately, little research has been conducted on this topic but, where such research is available, it provides clear and consistent evidence of the relevance of this principle for the self-concept.

Four points are worth noting in this regard. The first is that one cannot appreciate the significance of a specific self-concept component for global self-esteem if one fails to recognize the importance or centrality of that component to the individual. Empirical confirmation of this point appears in a study of high school juniors and seniors (Rosenberg, 1965). As an example, these students were asked how "likeable" they thought they were. As we would anticipate, those who considered themselves likeable were more likely to have high global self-esteem than those who believed they were not. But the strength of this relationship depended upon the importance attached to being likeable. Among those who *cared about* being likeable, the relationship of the self-estimate to global self-esteem was very strong, whereas among those to whom *this quality mattered little*, the relationship was much weaker.

The significance of self-values is particularly striking with regard to *negative* self-assessments. These adolescents were asked to judge themselves in terms of 16 traits or qualities. We will consider just those pupils who ranked themselves low in regard to these qualities, that is, they did *not* consider themselves likeable, or dependable, or intelligent, or conscientious, etc. How many of these people had low global self-esteem? The answer is that it *depended on how important each of these qualities was to the individual*. With regard to 15 of these 16 qualities, those who cared about the quality had lower self-esteem than those who considered the quality unimportant. Yet these people ranked themselves *the same way* with respect to the qualities in question.

In sum, to know that someone considers himself deficient with regard to a particular quality is plainly an inadequate indication of what he thinks of himself. We must also know how much he values this quality. If a particular component is vital to one's feeling of worth, then negative attitudes concerning it may be personally devastating, but if the component is trivial or insignificant, then the individual may blithely acknowledge inadequacy in that regard with scarcely a twinge of discomfort.

The second point is that the self-concept is less competitive than it might at first appear. This may sound like a strange message, first, because we have placed such a heavy emphasis on the importance of social comparison for self-assessment (involving the explicit assessment of the self in relation to others) and, second, because the competitiveness of American society is notorious (Murphy, 1974; Williams, 1951). What the principle of psychological centrality calls to attention, however, is that, to the extent that individuals focus their sense of worth on *different* self-components, the success of one person is not necessarily achieved at the expense of another.[11] It is thus entirely possible for each person to judge himself favorably by virtue of selecting his own criteria for judgment. Take four boys. One is a good scholar, the second a good athlete, the third very handsome, and the fourth a good musician. So long as each focuses on the quality at which he excels, each is superior to the rest. At the same time each may blithely acknowledge the superiority of the others with regard to qualities to which he himself is relatively indifferent. It is thus possible for each to emerge with a high level of self-respect and, indeed, mutual respect.

Much has been made of the high level of competition prevailing in capitalist societies characterized by a refined division of labor, and it is not our intent to dispute this claim. At the same time, the division of labor is such that each individual is encouraged to develop his own special area of expertise. Except insofar as the universal medium of money renders different realms of endeavor comparable, it is possible for people to accept and admire the skills of others in every realm of endeavor except their own with little offense to their self-esteem. This fact, too, makes noncompetitive societies less noncompetitive than they might at first seem. Every society has its own standards of excellence and judges its members accordingly. If the male skills valued in a society are farming, or spear throwing, or magic, then men will be assessed in those terms, with the brilliance of one casting a shadow on the other.

Furthermore, since complex societies allow achievement in diverse activities, and since they afford considerable (though not complete) leeway in the selection of self-values, one would expect their members to regard most highly those qualities at which they believe they excel. And the empirical data strikingly support this expectation. In the Rosenberg (1965) study of adolescents, with regard to every one of the 16 qualities under consideration, subjects who evaluated themselves favorably considered that characteristic more important personally than those rating themselves unfavorably. Thus, the individual strives to excel at that which he values and to value that at which he excels. Different characteristics become cardinal in different people's self-concepts, with the self-satisfaction felt by one person not necessarily diminishing the self-satisfaction of another. For a number of reasons, unfortunately, there are limits to the application of this principle (see chapter 11), so that some people do end up with low self-esteem. Nevertheless, because of psychological centrality, it is possible for more people to have high than low self-esteem.

Finally, it is important to consider psychological centrality in relation to self-concept change. There is considerable inconsistency in the literature concerning the difficulty of changing the self-concept. Some experimental psychologists and sociologists (e.g., Videbeck, 1960; Maehr, *et al.*, 1962; Webster and Sobieszek, 1974) appear to experience no difficulty whatever in changing the self-concept: some simple experimental stimulus brings about a transformation in how their subjects view themselves. On the other hand, many depth psychologists and psychiatrists report that the most intense and probing analytic efforts, carried out over an extended period of time, are futile to change the self-concept. This fact leaves us with the question: is it easy or hard to produce change in the self-concept?

The answer, of course, lies in whether the self-concept component under consideration is central to the individual's feeling of worth or is simply a peripheral and unstructured self-component. Consider the experimental study of Videbeck (1960) which has served as a springboard for a good deal of further research (for example, Maehr, *et al.*, 1962; Webster and Sobieszek, 1974). In this investigation, 30 subjects from introductory speech classes were asked to read six poems. A researcher introduced as a visiting "speech expert" arbitrarily rated half the subjects favorably and the other half unfavorably in terms of emotional tone, voice control, conveying meaning, and other skills. Those receiving approval raised their self-evaluations on these skills whereas those receiving disapproval lowered them. But since speech evaluations probably mattered little to these subjects,[12] it was not difficult to produce self-concept change under these circumstances.

Thus, if we are to deal with the thorny problem of self-concept change, it is clearly essential to take account of psychological centrality. An experimenter can easily convince us that we are poor connoisseurs of white burgundy, but can he as easily convince us that we are fascists or latent homosexuals? Will an intellectual as easily accept another's judgment of him as stupid as the other's judgment of him as lacking in neatness? Whether it is difficult or easy to change a self-concept component thus depends in large part on *how critical it is to the individual's system of self-values*. The person who has staked himself solidly on certain statuses or dispositions may resist, with all the resources at his disposal, any efforts to change these elements, for his very concept of self and feeling of self-worth rest on these foundations.

The four principles enunciated above are, we believe, essential to the understanding of self-concept formation and will be invoked consistently throughout this work to account for the observed empirical data. The first two principles—reflected appraisals and social comparisons—are more conspicuously social in the sense that the individual, either directly or indirectly, sees himself from the point of view of other people or compares himself to referent others or reference groups. The latter two—self-attribution and psychological centrality—appear more purely psychological but are heavily influenced by social factors. The individual may assess himself by observing his behavior or its outcomes, but such assessments can only be made in terms of the standards, criteria, or frames of reference provided by the culture. And, as far as psychological centrality is concerned, self-values are heavily influenced by the value system of a society, and by the system of social rewards and punishments which thrusts certain qualities into the center of concern while relegating others to the periphery. All four principles bear upon the way we see, wish to see, and present ourselves.

It is worth saying a word about the somewhat vague term "principle." In speaking of a principle, we have in mind a generalization, not a law. The principle, then, represents a mode of conceptualization that makes sense of empirical data but, being general and lacking refinement, does not hold under all conditions. Frequently, then, the principle will not explain the empirical data so much as serve as a springboard for explanation. If a sound principle is invoked to account for a phenomenon but fails to do so, it is not because the principle is wrong but because it is in need of refinement and specification. Much of part II of this work (chapters 3–7) focuses on the application of one or another of these principles to selected empirical issues, and on the specification of the conditions under which the principles hold.

## NOTES

1. Witness the opening lines of Edgar Allen Poe's (1938: 274) classic horror story "The Cask of Amontillado:" "The thousand injuries of Fortunato I had borne as I best could; but when he ventured upon insult, I vowed revenge."

2. Self-esteem, it should be noted, does not show a linear progression with severity of mental illness, as ordinarily conceived. Both Wylie (1961: 216) and Kaplan (1975: 171) report that the self-esteem of psychotics is as high as, or higher than, the self-esteem of neurotics.

3. Some of the Freudian defense mechanisms were not actually developed by Freud; the term "rationalization" was introduced by Ernest Jones.

4. This pattern, called the "egotistical pattern," has appeared repeatedly in the literature, although there are exceptions. Illustrations of the egotistical pattern can be found in Johnson, Feigenbaum, and Weiby (1964); Streufert and Streufert (1969); Eagly (1967); and Snyder *et al.* (1976). See Miller and Ross (1975) for a general discussion of this literature.

5. It should be pointed out that the fact that there is a consistent association between the individual's view of himself and others' views of him does not prove that he sees himself that way *because* of others' views; in principle, it is possible that he and other people independently draw the same conclusions about his self on the basis of the same objective evidence. But at least it can be said without fear of contradiction that the data are clearly consistent with the conclusion that the opinions other people hold of us importantly shape our self-definitions.

6. Since the entire area is rife with terminological inconsistency, it is necessary to be explicit about our own usage. The term "self-concept," of course, refers to the individual's idea of himself, and "perceived self" to his view of what others think of him, but there is no generally accepted term to designate what other people *actually* think of the individual. Miyamoto and Dornbusch (1956) use the term "actual self," Sherwood (1965) "objective public esteem," and Rosenberg (1973a), "accorded self." Unfortunately, the first is misleading, the second, awkward, and the third, unfamiliar. Hence, despite some occasional confusion of meaning, we have opted for James' (1890) term "social self." "A man's social self," according to James (1890: 293) "is the recognition which he gets from his mates. . . . Properly speaking, a man has as many social selves as there are individuals who recognize him and carry an image of him in their mind. . . . But as the individuals who carry the image fall naturally into classes, we may practically say that he has as many different social selves as there are distinct *groups* of persons about whose opinion he cares. . . . A man's *fame*, good or bad, and his honor or dishonor, are names for one of his social selves." In speaking of the social self, then, we are referring to something external to the individual—the actual attitudes of particular people or groups of people toward him.

7. Precisely how people draw conclusions about others' attitudes toward themselves remains far from clear. As Backman and Secord (1962: 321) point out: "Although in our society O [the other] rarely communicates directly to S [the subject] his estimate of S's attributes, S presumably uses indirect signs or cues based on the behavior of the other to infer what O thinks of him."

8. This distinction parallels that of Kelley's (1952) two functions of reference groups: the comparison function and the normative function.

9. As we shall indicate later, this finding appears consistently, although the relationship tends to be only a moderate one.

10. The four criteria of internal validity, according to Kelley (1967: 1971), are distinctiveness, consistency over time, consistency over modality, and consensus.

11. The story is told of a famous violinist who accompanied the distinguished pianist, Vladimir Horowitz, to a recital by the then young violinist, Jascha Heifetz. At the intermission, Horowitz's companion mopped his brow and remarked: "Awfully hot in here, isn't it?" "Not for pianists," replied Horowitz.

12. Videbeck (1960), it should be stressed, was clearly aware that the likelihood of changing self-concept components depended on a number of conditions or circumstances.

# PART II

# Social Determinants of

# the Self-Concept

IN CHAPTER 2, four general principles were advanced to account for differences in self-esteem: reflected appraisals, social comparison processes, self-attribution, and psychological centrality. These principles are, we believe, general and sound, applying equally to children, adolescents, and adults. Yet, where they have implicitly or explicitly been invoked to predict the relationships between social or interpersonal factors and self-esteem, they have frequently resulted either in flatly erroneous or else very inexact predictions. The challenge of the next five chapters is to understand why this is so. In the course of attempting to make sense of certain of these puzzling findings, we hope to be able to elucidate some of the processes whereby social factors shape the self-concept.

Chapters 3–7 are based primarily on the analysis of two bodies of empirical data. The first is a sample of 5,024 public high school juniors and seniors in New York State, the second a sample of 1,988 school pupils from grades 3–12 in the Baltimore City public schools. (Details will be presented in due course.) The conclusions to be drawn, then, apply to young people, specifically from the period of middle childhood to late adolescence. The aim of part II is to understand how social forces—interpersonal, contextual, and structural—influence the self-concepts of these preadults. For comparative purposes, these data are supplemented in chapter 5 with parallel data drawn from a sample of 2,300 adults in the Chicago Urbanized Area.

In attempting to shed light on how social life forms and shapes the self-concepts of young people, our strategy is to proceed from direct, immediate, influences on the self-concept to broad factors in the social structure rooted in complex historical, political, and technological forces. Chapter 3, then, starts with a consideration of the impact of direct, face-to-face interaction—especially with "significant others"—on children's and adolescents' self-concepts. Chapter 4 turns to the bearing of the immediate social context, particularly the contexts of neighborhood and school, on selected aspects of the self-concept. Finally, chapters 5–7 center on the significance of the individual's location in the broader social structure for his ultimate self-picture. Chapter 5 considers the general system of social stratification that characterizes American society. The key question considered is whether the importance of social class position for the child's self-concept is necessarily the same as for the adult's. Chapters 6–7 take up the question of racial and religious stratification in relation to the views young people hold of themselves. Social forces influence children's self-concepts but these effects, as we shall see, are not always those one would expect.

# 3

# Which Significant Others?

IT MAY generally be asserted that, whenever important sociological theories are actually brought into direct contact with empirical reality, they seldom remain "whole." Rather, they experience
one of two fates: either they are attacked and abandoned, or they
undergo progressive refinement and specification. The principle of
reflected appraisals, we shall see throughout this book, illustrates this
process of increasing refinement.

Consider Cooley's "looking-glass" principle—the view that we
come to see ourselves as we believe others see us. This principle,
as noted in the previous chapter, has been dramatically supported by
empirical research; there is a strong and definite relationship between the "perceived self" and the individual's own picture of what
he is actually like.

Still, there are cases in which this principle does not hold. Some
people who believe others think well of them have negative self-
attitudes, and vice versa. The purpose of this chapter is to suggest
that one reason such discrepancies exist is that the attitude *toward*
the other, as well as the attitude *of* the other, may play a role in
self-concept development. In other words, *not all significant others
are equally significant,*[1] and those who are more significant have
greater influence on our self-concepts.

That particular others are not equally significant has often been noted (Rose, 1962; Hughes, 1962; Manis, 1955; Stryker, 1962; Sullivan, 1953),[2] and the relevance of degree of significance has been documented (Manis, 1955). In most discussions, however, putative significant others are involved; that is, the investigator assumes the other is significant. For example, Sullivan (1953) assumes that parents are significant others, Denzin (1966) that professors, classmates and spouses are significant others, Manis (1955) that friends are significant others, and the like. These assumptions are reasonable and, on the whole, probably sound. But significance is in the eye of the beholder; ultimately, he alone can determine whether or not a particular other is significant to him. One question, then, is whether *putative* significant others are in fact *real* significant others. If not, what is the effect of this differential degree of significance on the self-concept? Furthermore, what are some of the factors affecting this differential degree of significance?

The second question is: what do we mean by "significant?" Different people may be significant to us in different respects and for different reasons. In the present discussion, we shall distinguish two closely related aspects of significance: *valuation* and *credibility*. The first deals with whether the opinion of the other is valued, the second with whether it is respected.

When we speak of a valued other, we refer to someone whose favorable opinion we strongly desire. This is a person whose opinion we care about, whose opinion makes a difference to us. Respect, on the other hand, involves confidence in the judgment of the other. A child's opinion of himself may be strongly influenced by what he believes his teacher thinks of him, not necessarily because he wants the teacher to like him but because he attributes to the teacher superior knowledge and insight. Conversely, a child might want other children to like him without having faith in their ability to assess his true worth.

Finally, how does the individual's location in the social structure affect his attribution of significance? In other words, how may his statuses, with their associated "role-sets," influence the range of others with whom he interacts and the quality of these interactions?

## SAMPLE

The data to be analyzed in this chapter—indeed, the chief data set used throughout the remainder of this book—is a sample drawn from the population of public school pupils of Baltimore City.

A cluster sample was employed in this study. Each school in Baltimore City was initially stratified on the basis of two variables: (1) the proportion of nonwhite students, and (2) the median income of its census tract. Twenty-five schools falling into the appropriate intervals were randomly selected with a probability proportionate to size. From each school, 105 children were selected from the central records by random procedures.

Some children had withdrawn from school subsequent to the compilation of the central records and were no longer available. However, we were able to interview 1,917 children—that is, 79.2 percent of the sample children still registered in the schools, or 73 percent of all the children originally drawn from the central records. Reflecting the 1968 population, the sample is 63 percent black and somewhat skewed toward the working class.[3]

Each subject was interviewed directly after classes at his own school. For the elementary school children, objective background information was collected from the parents. Parents were contacted either by a 5 to 10 minute telephone interview or, when there was no telephone, by home interview. Almost all parents were extremely cooperative and in only 60 cases were we unable to locate the parent or conduct the interview.

## Interpersonal Significance

In order to investigate the impact of various significant others on the self-concept, three concepts must be clarified and measured:

(1) *Global self-esteem*: This is the individual's global positive or negative attitude toward himself. In our usage, the individual with high self-esteem considers himself a person of worth, though he does not necessarily believe he is superior to others. Low self-esteem, on the other hand, implies self-rejection, self-dissatisfaction, or self-contempt. In the Baltimore study, self-esteem is measured by a six-item Guttman scale (reproducibility = 90 percent, scalability = 68 percent).[4] (Example: "Everybody has some things about him which are good and some things about him which are bad. Are most of the things about you good, bad, or are both about the same?")

(2) *Perceived self*: The individual's concept of how others judge

and evaluate him is his perceived self. But since different people may hold different views of him, it is apparent that everyone has a multiplicity of perceived selves. In this study, we attempted to obtain information on most of the perceived selves which might reasonably be considered significant to the child—that is, putative significant others. Several different indicators were employed. In one set of questions, we asked our respondents: "Would you say your mother thinks you are a wonderful person, a pretty nice person, a little bit of a nice person, or not such a nice person?" The same inquiry was made of the opinions of "your teachers," of "the kids in your class," and of "your father."

At another point, the respondents were asked, "How much do boys like you?" and "How much do girls like you?" [5] Although we did not ask the respondents what their siblings thought of them, we did pose several questions dealing with how well the respondent got along with his brothers and sisters. In lieu of more direct evidence, we shall assume that, roughly speaking, the child who feels that he gets along well with his siblings is more likely to believe that they have a favorable opinion of him than the one who does not.

(3) *Interpersonal valuation*: This concept refers to concern with the other's opinion of oneself. Just as the individual may attach varying degrees of importance to one or another *characteristic* as a criterion for self-judgment, so he may attach great or little importance to one or another *person* as a basis for self-judgment.

The interviewers introduced this topic by saying:

We may care what some people think of us and not care too much what other people think. I am going to mention some people and I would like you to tell me how much you care about what they think of you. First, how much do you care about what your MOTHER thinks of you? Do you care very much, some, not very much, or not at all?

The same question was asked with regard to the child's father, teachers, "most kids" in his class, friends, and brothers and sisters.

## VALUATION

The empirical data clearly support two of the points enunciated in our introduction to this chapter. The first—that not all significant others are equally significant—is shown in the fact that, whereas 84 percent of the respondents care "very much" what their mothers think of them, only 30 percent assign equal importance to "kids in your class." [6] The second point, that *putative* significant others are not necessarily *real* significant others, is also supported by the data. In fact, fully one-sixth of the children even denied that they cared "very much" about their *mothers'* opinions of them and one-fourth attributed equally low significance to their fathers; with regard to classmates, the proportion reached seven out of ten.

The key question then becomes: is the relationship between the perceived self and self-esteem stronger if the other is highly significant than if he is not? The data relevant to this question appear in Table 3–1. Since seven perceived selves and putative significant others are involved, Table 3–1 requires some explanation. For example, line 1 of Table 3–1 deals with whether respondents who believe their mothers think they are "wonderful" are more likely to have high self-esteem than those who believe their mothers consider them "not such a nice person." The data show that this relationship is moderate (gamma = .1759) and statistically significant ($p < .05$) [7] if the respondent cares "very much" or "pretty much" what his mother thinks of him but weak (gamma = –.0456) and not statistically significant if he cares "little" or "not at all" about her opinion of him. (A description of the conventions regarding statistical significance used in this work is presented in footnote 7). The remainder of the comparisons deal with different others—father, teachers, classmates, boys, girls, and brothers or sisters.

The results are clear and generally consistent. With regard to six of these seven putative significant others, the association between the perceived self and self-esteem is stronger among those who care what these significant others think of them than among those who do not; the remaining case is ambiguous.

We thus see that the perceived self will chiefly be associated with self-esteem *if the particular significant other is really significant* to the individual—that is, if the other's favorable opinion is strongly valued. It is not only what we believe others think of us, but what each of them means to us personally, that affects self-esteem.

But if the individual is in a position to decide for himself which other people are or are not significant, how will he make this decision? It will come as no surprise to learn that he is likely to attribute significance in such a way as to maximize his self-esteem. Specifically, he is likely to conclude that the people whose opinions really matter to him are those who, in his view, think highly of him; conversely, he tends to be relatively unconcerned with the views of his detractors.

The data appear in Table 3–2. In each case, the child who has come

## TABLE 3-1.

*Perceived Self and Self-Esteem by Concern with Opinions of Various Putative Significant Others*

| | Interpersonal Valuation | | | | | |
| | Respondent cares "very" much or "pretty" much what significant other thinks of him | | | Respondent cares "little" or "not at all" what significant other thinks of him | | |
| | Respondent believes other's view of him is . . . | | | | | |
| | Favor-able | Inter-mediate | Unfavor-able | Favor-able | Inter-mediate | Unfavor-able |
|---|---|---|---|---|---|---|
| **Mother**[a] | | | | | | |
| High Self-Esteem | 48% | 41% | 24% | 13% | 36% | 39% |
| N = 100% | (488) | (1,071) | (163) | (16) | (42) | (31) |
| Favorable-unfavorable | | | | | | |
| % difference | | 24 | | | −26 | |
| Gamma | | *0.1759 | | | −0.0455 | |
| **Father**[a] | | | | | | |
| High Self-Esteem | 48% | 39% | 34% | 35% | 40% | 34% |
| N = 100% | (527) | (773) | (110) | (17) | (45) | (56) |
| Favorable-unfavorable | | | | | | |
| % difference | | 14 | | | 1 | |
| Gamma | | *0.1362 | | | −0.0177 | |
| **Teachers**[a] | | | | | | |
| High Self-Esteem | 55% | 42% | 31% | 73% | 37% | 34% |
| N = 100% | (174) | (1,236) | (196) | (11) | (98) | (70) |
| Favorable-unfavorable | | | | | | |
| % difference | | 24 | | | 39 | |
| Gamma | | *0.2577 | | | *0.2001 | |
| **Kids in Class**[a] | | | | | | |
| High Self-Esteem | 55% | 42% | 26% | 25% | 45% | 33% |
| N = 100% | (123) | (1,022) | (129) | (24) | (322) | (167) |
| Favorable-unfavorable | | | | | | |
| % difference | | 29 | | | −8 | |
| Gamma | | *0.2174 | | | *0.1441 | |
| **Boys**[b] | | | | | | |
| High Self-Esteem | 48% | 38% | 27% | 51% | 44% | 40% |
| N = 100% | (236) | (575) | (67) | (113) | (434) | (258) |
| Favorable-unfavorable | | | | | | |
| % difference | | 21 | | | 9 | |
| Gamma | | *0.2293 | | | 0.1314 | |
| **Girls**[b] | | | | | | |
| High Self-Esteem | 40% | 43% | 22% | 43% | 47% | 41% |
| N = 100% | (418) | (717) | (91) | (70) | (250) | (194) |
| Favorable-unfavorable | | | | | | |
| % difference | | 18 | | | 2 | |
| Gamma | | *0.0743 | | | 0.0928 | |
| **Siblings**[c] | | | | | | |
| High Self-Esteem | 49% | 40% | 28% | 40% | 30% | 32% |
| N = 100% | (584) | (698) | (76) | (40) | (138) | (54) |
| Favorable-unfavorable | | | | | | |
| % difference | | 21 | | | 8 | |
| Gamma | | *0.1665 | | | 0.0648 | |

[a] Mother, father, teachers, kids in class think you are "wonderful" (favorable), "pretty nice" (intermediate), or "a little nice" or "not so nice" (unfavorable).

[b] Would you say that boys, girls like you "very much" (favorable), "pretty much" (intermediate), or "a little" or "not at all" (unfavorable).

[c] Get along with brothers and sisters "very well" (favorable), "pretty well" (intermediate), or "not so well" (unfavorable).

*Signifies that the table upon which the gamma is based is significant, according to the chi square calculation. For a description of the conventions regarding statistical significance used throughout this work, see footnote 7, p. 98.

to the conclusion that one of his significant others—mother, father, teachers, or classmates—thinks poorly of him is much more likely to decide that he "doesn't care" what they think; and if, indeed, he is successful in internalizing this valuation, then he can very effectively protect his self-esteem. The child is thus not simply a passive object, a lump of clay totally molded by his interpersonal environment; rather, he reacts to this environment in a selective way, a selectivity protective of his self-esteem. This selectivity mechanism, to be sure, is limited by reality. As Table 3–2 shows, even children who believe their mothers think ill of them rarely deny that their mothers' opinions are important to them; the mother-child role relationship is so powerful that it is not easily overcome by selective valuation. With regard to fathers, teachers, or classmates, however, the arrangement of one's interpersonal value hierarchy is very strongly directed by the desire to protect one's self-esteem. For example, only 38 percent of the respondents who believed their teachers thought poorly of them said they cared very much about their teachers' opinions, compared with 81 percent of those who attributed favorable attitudes to their teachers. For fathers, the corresponding figures were 41 percent and 90 percent, and for "kids in your class," 13 percent and 52 percent.

When we speak of the individual "deciding" that he does or does not care what specific others think of him, of course, we do not mean

**TABLE 3-2.**

*Child's Perception of How Significant Others Rate Him
and Degree to Which He Values Their Good Opinion*

| Care "very much" what . . . | Respondent believes significant other considers him . . . | | |
| --- | --- | --- | --- |
| | Favorable | Intermediate | Unfavorable |
| *Mother thinks of him[a] | 91% | 85% | 64% |
| N = 100% | (525) | (1,155) | (208) |
| *Father thinks of him[a] | 90% | 77% | 41% |
| N = 100% | (573) | (842) | (174) |
| *Teachers think of him[a] | 81% | 58% | 38% |
| N = 100% | (199) | (1,386) | (278) |
| *Kids in class think of him[a] | 52% | 31% | 13% |
| N = 100% | (157) | (1,400) | (305) |

[a] Mother, father, teachers, kids in class think you are "wonderful" (favorable), "pretty nice" (intermediate), or "a little nice" or "not so nice" (unfavorable).

*See footnote 7, p. 98.

to imply that this is a conscious, rational decision. The adjustment is probably a gradual and unconscious one in which a shift in the significance attributed to particular others may occur in scarcely perceptible degrees. The more the other person criticizes or disapproves of him, the more will the individual try to shrug it off, discount their judgment, withdraw affect from them. Although he will not be entirely successful, the inclination is there. In the long run, then, he is likely to end up caring most about the opinions of those who, in his view, think well of him.

CREDIBILITY

It is a familiar dictum of communications research that the effectiveness of a message depends not only on *what* is said but on *who* says it. Since the self-concept may be viewed as an attitude toward an object, it would follow that the prestige or respect accorded the other person would equally influence the degree to which this perceived self bears upon our self-concept. Like valuation, respect is in the eye of the beholder, and the individual himself must designate the people in whom he has faith. The question is: Is the association between what we believe others think of us and what we think of ourselves stronger if we have high faith in others' judgments (high source credibility) than if we have low faith in them (low source credibility)?

Source credibility was assessed in two ways. After being asked what their parents, teachers, and best friend would say about them to others, respondents were then instructed: "You told me a moment ago what your parents would say about you. Would your parents be mostly right, somewhat right, or mostly wrong in what they would say about you?" Similarly, the respondents were asked whether their "teachers" and their "best friend" would be mostly right, somewhat right, or mostly wrong.

The other question which served as an indicator of source credibility is the following: "Does your MOTHER really know what you are like deep down inside?" The same question was asked for the father, any brothers or sisters, friends, and teachers.

Tables 3–3 and 3–4 clearly indicate that the relationship between the perceived self and self-esteem is stronger if the other person is trusted than if he is not. Table 3–3 presents the relationship between what the child believes his parents, teachers, and best friend think of him and his personal self-esteem among those who believe that these

**TABLE 3-3.**

*Attitudes Attributed to Significant Others and Self-Esteem, by Trust in Judgment of Significant Other*

| | Believe parents' description of respondent would be . . . | | | | | | | | |
|---|---|---|---|---|---|---|---|---|---|
| | Mostly right | | | Somewhat right | | | Mostly wrong | | |
| | How parents would describe respondent to others | | | | | | | | |
| | Favor-ably | Mixed, Neutral | Unfavor-ably | Favor-ably | Mixed, Neutral | Unfavor-ably | Favor-ably | Mixed, Neutral | Unfavor-ably |
| High self-esteem | 51% | 38% | 24% | 36% | 34% | 26% | 23% | 43% | 41% |
| N = 100% | (713) | (261) | (104) | (261) | (141) | (57) | (22) | (23) | (34) |
| Favorable-unfavorable | | | | | | | | | |
| % difference | | 27 | | | 10 | | | −18 | |
| Gamma | | *0.3118 | | | 0.0300 | | | −0.2038 | |

| | Believe teachers' description of respondent would be . . . | | | | | | | | |
|---|---|---|---|---|---|---|---|---|---|
| | Mostly right | | | Somewhat right | | | Mostly wrong | | |
| | How teachers would describe respondent to others | | | | | | | | |
| | Favor-ably | Mixed, Neutral | Unfavor-ably | Favor-ably | Mixed, Neutral | Unfavor-ably | Favor-ably | Mixed, Neutral | Unfavor-ably |
| High self-esteem | 49% | 36% | 36% | 40% | 32% | 27% | 37% | 37% | 31% |
| N = 100% | (660) | (256) | (92) | (305) | (151) | (70) | (30) | (19) | (36) |
| Favorable-unfavorable | | | | | | | | | |
| % difference | | 13 | | | 13 | | | 6 | |
| Gamma | | *0.2420 | | | 0.1782 | | | −0.0122 | |

| | Believe best friends' description of respondent would be . . . | | | | | | | | |
|---|---|---|---|---|---|---|---|---|---|
| | Mostly right | | | Somewhat right | | | Mostly wrong | | |
| | How best friend would describe respondent to others | | | | | | | | |
| | Favor-ably | Mixed, Neutral | Unfavor-ably | Favor-ably | Mixed, Neutral | Unfavor-ably | Favor-ably | Mixed, Neutral | Unfavor-ably |
| High self-esteem | 46% | 35% | 28% | 36% | 33% | 26% | 34% | 30% | 24% |
| N = 100% | (840) | (118) | (32) | (345) | (73) | (23) | (73) | (37) | (29) |
| Favorable-unfavorable | | | | | | | | | |
| % difference | | 18 | | | 10 | | | 10 | |
| Gamma | | *0.1952 | | | 0.1256 | | | 0.0934 | |

*See footnote 7, p. 98.

others are mostly right, somewhat right, or mostly wrong in their judgments of him. The results are clear and consistent. In each case, the stronger the belief that the significant other is right about us, the greater the apparent effect of that perceived judgment on our self-esteem. For example, consider those who trust their teachers' judgments of them (believe teachers "mostly right"). If they believe their teachers view them favorably, then their self-esteem is high, and if they think their teachers view them negatively, then their self-esteem is lower (gamma = .2420). But if they believe their teachers do *not* know what they are like, then the self-esteem of those who think their teachers view them unfavorably differs little from those who believe their teachers view them favorably (gamma = –.0122).

Table 3-4, dealing with the "deep down inside" questions, points

## TABLE 3-4.

*Perceived Self and Self-Esteem, by Credibility of Significant Other*

| | Knows you deep down inside? | | | | | |
| | Yes | | | No | | |
| | Attitude perceived as . . . [a] | | | | | |
| | Favor-able | Inter-mediate | Unfavor-able | Favor-able | Inter-mediate | Unfavor-able |
|---|---|---|---|---|---|---|
| **Mother** | | | | | | |
| High self-esteem | 50% | 44% | 23% | 37% | 36% | 31% |
| N = 100% | (361) | (708) | (101) | (115) | (364) | (82) |
| Favorable-unfavorable | | | | | | |
| % difference | | 27 | | | 6 | |
| Gamma | | *0.2213 | | | 0.0126 | |
| **Father** | | | | | | |
| High self-esteem | 50% | 42% | 31% | 43% | 36% | 34% |
| N = 100% | (319) | (361) | (36) | (196) | (423) | (120) |
| Favorable-unfavorable | | | | | | |
| % difference | | 19 | | | 9 | |
| Gamma | | *0.1715 | | | 0.0594 | |
| **Teacher** | | | | | | |
| High self-esteem | 62% | 41% | 32% | 45% | 41% | 30% |
| N = 100% | (103) | (425) | (57) | (69) | (850) | (194) |
| Favorable-unfavorable | | | | | | |
| % difference | | 30 | | | 15 | |
| Gamma | | *0.3396 | | | *0.1755 | |

[a] Mother, father, teachers think you are "wonderful" (favorable), "pretty nice" (inter-mediate), or "a little nice" or "not so nice" (unfavorable).
*See footnote 7, p. 98.

to similar conclusions. Those who trust the judgment and insight of the significant other are more likely to be affected by what they perceive the other thinks of them than those who do not. Although there is some variation in degree, the results are strong and consistent for mothers, fathers, and teachers. It is thus not simply what significant others think of us but also how much we trust their judgment that bears upon our self-concepts.

SOURCES OF DIFFERENTIAL SIGNIFICANCE

What accounts for our propensity to accord high respect to certain people's judgments of us, little respect to others? Many factors are involved, ranging from personal motivation to broader forces of the social structure. Four of these factors will be considered: self-concept motives, areas of expertise, degree of consensus, and role-set.

*Self-concept Motives.* Like valuation, interpersonal credulity is also a matter of motivation—the motivation to protect and enhance the self-concept. This desire enters through what might be called the mechanism of "selective credulity," that is, having stronger faith in the judgments of those who appreciate our merits than in those more alert to our shortcomings. This is a familiar aspect of human experience; our admiration for another's wit and wisdom often rockets when we learn that he holds us in high regard while we are flabbergasted at the dimwittedness of our critics. Such selective credulity is as powerful among children as among the rest of us. Table 3-5 shows that, in each of six comparisons, those who attribute more favorable remarks or opinions to others are more likely to trust the opinions of these others than those who infer less favorable attitudes. If we believe that particular others think well of us, then these are the people who, in our opinion, truly understand what we are like.

*Areas of Expertise.* One reason that the concept of interpersonal centrality is so intricate empirically is that some people are assumed to be experts with reference to certain aspects of the self, other people with reference to other aspects.

A pointed example is Denzin's (1966) study of the significant others of a college population. Using the work of Kuhn (1964) and of Mead (1934) as a springboard, Denzin distinguished between "role-specific significant others," (characteristically some aspect of the individual's role-set) and "orientational others" (based on relationships which cross-cut a variety of situations). In order to determine role-specific significant others, subjects were asked: "Would you

**TABLE 3-5.**

*Child's Perception of What Significant Others Would Say
About Him and Acceptance of Their Assessment*

| | Adult description of child[a] | | |
| --- | --- | --- | --- |
| | Favorable | Mixed, Neutral | Unfavorable |
| *Believe *parents* would be mostly right | 71% | 62% | 55% |
| N = 100% | (1,031) | (449) | (203) |
| *Believe *teachers* would be mostly right | 66% | 60% | 47% |
| N = 100% | (1,029) | (446) | (209) |
| *Believe *best friend* would be mostly right | 67% | 52% | 39% |
| N = 100% | (1,301) | (239) | (87) |

| | Adult evaluation of child[b] | | |
| --- | --- | --- | --- |
| | Favorable | Mixed, Neutral | Unfavorable |
| **Mother* knows you deep down inside | 76% | 66% | 55% |
| N = 100% | (495) | (1,112) | (197) |
| **Father* knows you deep down inside | 61% | 46% | 23% |
| N = 100% | (539) | (806) | (164) |
| **Teachers* know you deep down inside | 59% | 33% | 23% |
| N = 100% | (186) | (1,321) | (263) |

[a] Coded responses to question: "Let's pretend your parents (teachers, best friend) wanted to tell someone all about you. What would they say? What type of person would they say you were?"

[b] Mother, father, teachers think you are "wonderful" (favorable), "pretty nice" (intermediate), or "a little nice" or "not so nice" (unfavorable).

*See footnote 7, p. 98.

please give me a list of those persons or groups of people whose evaluation of you as a *student* on the _____ campus concerns you the most." "Orientational significant others" were assessed by the question: "Would you please give me a list of those persons or groups of people whose evaluation of you as a *person* concerns you the most."

Based on a small sample of subjects, the results showed that, with regard to evaluation of the self as a *student,* respondents were most likely to mention faculty, with friends and family members following in that order. But when asked about the views of others toward the self as a *person,* friends were most frequently mentioned first, followed by family members, with faculty in third place. Interpersonal valuation thus depends on which aspect of the self is under consideration.

*Degree of Consensus.* Statements to the effect that we see ourselves as others see us are most clearly recognized as oversimplifications when we realize that different others may hold different views of us in the same or different respects. How can our self-attitudes be direct reflections of others' attitudes if different people see us differently? One factor that will affect us, then, is whether or not we believe these different others *agree* with one another about what we are like.

A study by Backman, Secord, and Pierce (1963) aptly illustrates this point. In this study, the subject was asked to write down the names of five close friends or relatives whose opinions he valued, and to indicate how he believed each would describe him in terms of certain "need statements" (perceived selves). By means of an ingenious manipulation, the investigators attempted to change the subject's views of two self-concept components—one on which the subject believed consensus to be high, the other on which he believed consensus to be low. If low, the need component proved easy to change, but if high, it was much more difficult to change. These results indicate that the self-concept will be more closely related to the perceived self if we believe others agree about what we are like than if we think they disagree.[8]

*Role-Set.* Interpersonal significance, of course, depends largely on the people with whom we interact. But such interaction is not random; to a considerable extent it is determined by social structural forces. Specifically, the others with whom we interact most extensively and intensively are determined by our "role-set." The term role-set was introduced by Merton (1957:369) to refer to the "complement of role relationships which persons have by virtue of occupying a particular social status." For example, the status of public school teacher involves patterned relationships with a number of others—pupils, colleagues, the school principal, the superintendent, the Board of Education, Parent-Teacher Associations, professional teachers' organizations, and so on. These patterned interactions are structured by the status, that is, the socially-defined position (Linton, 1936). Needless to say, each individual occupies a number of statuses.

The status of child in our society also has its characteristic role-set. What is important in the present context is the degree to which this particular *role-set virtually exhausts the individual's significant interactions*—with mother, father, siblings, teachers, friends, and classmates. For the child, this particular role-set is a primary determinant of the available pool of people who can potentially represent significant others to him.

The upshot is that almost all categories of children in America—

whatever their sex, race, or class—attribute significance to the same people and in much the same order. For example, we computed a rank correlation coefficient (Spearman rho) of interpersonal significance for various groups. In other words, if a particular group of children was most likely to say that they cared "very much" what their mother thought of them, then mother was ranked first for this group, and so on down the line. With regard to valuation the various social groups show a striking similarity: for sex, rho = .94; for race, .83; for age,[9] .92; and for socio-economic status,[10] .97. Otherwise expressed, whatever the child's sex, race, age, or socioeconomic status, the mother is most likely to be ranked as highly significant, followed by father, brothers and sisters, teachers, friends, and kids in your class, in that order. The role-set of children in different social categories is so broadly similar that the same categories of people tend to enter their lives as valued others. This is, however, less true with regard to credibility (know what you are like deep down inside). The Spearman rho for sex = .60; for race, .70; for age, .44; and for social class, .94.

## Summary and Discussion

One expression of the principle of reflected appraisals is Mead's (1934:68) observation that "we are more or less unconsciously seeing ourselves as others see us." By and large the evidence has borne out this assertion. The question is: why is it only "more or less?" One reason, of course, is that people do not "take the role of the other" with unerring accuracy; no human being can ever completely enter into the mind of another. Hence, as a proximal influence, it is the "perceived self"—what we *think* others think of us—that affects our self-attitudes. Mead's statement might be amended to read: "We are more or less unconsciously seeing ourselves as we *think* others see us."

But even the perceived self, though a more effective predictor of self-concept than the social self, is still imperfect. One reason, this chapter has suggested, is that not all "others" are equal. If we are deeply concerned about particular others' opinions of us, or if we trust their judgment, then our views of what they think of us will clearly be related to what we think of ourselves; but if we do not value their good will or trust their judgment, then the effect on our self-concept is almost invariably smaller. We can thus further amend Mead's statement to read: "We are more or less unconsciously seeing ourselves as we think others who are important to us and whose opinion we trust see us." The statement is not inconsistent with Mead's theory; it is only more precise.

That we see ourselves from the viewpoints of others and are heavily influenced by what they think of us—the principle of reflected appraisals—is a fundamental and undeniable axiom of the symbolic interactionist position. The scientific task in dealing with such truths is neither to challenge them (which would be futile) nor to confirm them (which would be trivial), but to refine them, specifying the conditions under which they operate to a greater or lesser degree. Several specifications have been presented in this chapter and additional ones will be presented at various points throughout this work. Hopefully, such progressive refinements improve the theory's ability to account for empirical reality in specific instances as well as in general.

## NOTES

1. Actually, Denzin (1966) points out that it was Harry Stack Sullivan, not Mead, who coined the term "significant other." Many sociologists have attributed the expression to Mead partly because his terms "significant symbol," "particular others," and "generalized other" are so well known, and partly because he clearly recognized the concept without using the term. We shall continue to counterpose "significant other" and "generalized other" in our discussion of Mead because these terms best communicate the ideas under discussion. Denzin (1966: 298) notes, however, that "There is some doubt whether Sullivan intended his term 'significant other' to refer to those responsible for socializing the actor or to all those persons the actor holds in high esteem. Recent usage has tended to employ the latter interpretation."

2. The point has been aptly expressed by Kinch (1967: 263): "The more important the individual perceives the contact between himself and others to be, the more likely it is that the individual's perceptions of the responses of the others will be used in defining his self-image."

3. One school—a combined elementary and junior high school—was entered twice in the total population of schools and, by chance, was selected in both categories. It was not practicable to double the sample size of this school; hence, the responses of these thirty-five elementary school children and those of the thirty-six junior high school

children were doubled in weight to better represent the total population. In our analysis, we have thus treated our sample as 1,988 children.

There is evidence to indicate that the sample is representative of the population from which it is drawn (that is, the public school population of Baltimore City from grades 3 through 12 in 1968). A comparison of sample data with population data shows an impressive similarity in terms of race, grade level, size of school, racial context of school, and median family income of census tract of school. For example, the population is 64.1 percent black, the sample 63.6 percent black. The sole deviation is some overrepresentation of elementary, and underrepresentation of senior high, pupils. (For details, see Rosenberg and Simmons, 1972: Appendix A).

4. For a description of the scale items and a discussion of the reliability and validity of the measure, see Appendix A-2. This scale is termed the "Baltimore Self-Esteem scale" or "Rosenberg-Simmons Self-Esteem Scale" (RSSE) throughout this work.

5. See Appendix B-1 for specific wording of perceived self indicators.

6. Brookover, Thomas, and Paterson (1964) also showed that parents were cited as significant others considerably more often than peers (among seventh graders).

7. The .05 level of significance is employed throughout this report, and is represented by an asterisk. The following conventions are used throughout this work:

1) If the independent variable is a dichotomy, the significance of difference between two proportions is computed. An asterisk between the *two first-row percentages* indicates that the difference is significant at the .05 level; the absence of an asterisk indicates it is not. Significance is based on the two-tailed test.

2) If the independent variable is a polytomy, significance is measured by chi square. An asterisk before the *table title* indicates that the zero-order relationship is significant. If the table is stratified according to the categories of a test factor, an asterisk before the *test factor category title* indicates that the relationship is significant.

3) If both variables are at the ordinal level of measurement and it is desirable to present summary statistics (especially for comparative purposes), the statistic gamma is employed. An asterisk before the *gamma* indicates that the table on which the gamma is based is statistically significant at the .05 level, according to the chi square test.

4) Although we lack interval level data, there are occasions when it has been necessary to use means and Pearsons *r*'s. An asterisk before the *r* indicates that it is statistically significant at the .05 level.

5) Significance levels are not computed for first or higher order associations, whatever method of control is employed—subgroup classification, test factor standardization, partial gamma, or partial correlation.

8. A study by Kinch (1963), involving a situational manipulation of the self-concept, affords another vivid illustration of this point.

9. Based on average intercorrelation of ranks of three age categories: 8–11, 12–14, and 15 or over.

10. Based on average intercorrelation of ranks of four social class categories. Social class has been computed according to the Hollingshead scale; because of small totals, classes I and II were combined.

# 4

# Contextual Dissonance:
# Self-Concept Effects

IF the discussion of interpersonal significance is based chiefly on the principle of reflected appraisals, the discussion of contextual dissonance rests fundamentally on the social comparison principle. The reason these principles are differentially relevant is that interpersonal significance is primarily concerned with immediate face-to-face relations whereas the social context focuses attention on the *groups* directly surrounding us, and our relations to them.

That the social similarity or dissimilarity of the individual to those around him may affect his experience, and in consequence his self-concept, is obvious. For example, the experience of being black when mostly surrounded by blacks is plainly different from being black in a white environment. Similarly, the meaning of being Jewish in a gentile neighborhood is not the same as being Jewish in a Jewish neighborhood. The difference reflects the consonance or dissonance of the environment for the individual.[1] Although Hessler *et al.* (1971) conceptualize the environment as dissonant or consonant in itself, our own approach is to view the context as dissonant or consonant for a given individual.[2]

Despite its general neglect, contextual dissonance or consonance is a quintessentially sociological phenomenon, directly concerning the relationship of the individual to his environment. This fact is apparent in the diversity of terms employed by sociologists to refer to various aspects of the relationship between the individual's characteristics and those prevalent in his environment. Consider such time-honored terms as "minority group," "segregation and integration," "marginal man" (Stonequist, 1937; Park, 1928), "pariah people" (Weber, 1952), and the "stranger" (Simmel, 1950). If the term "minority group" has any logical meaning, it refers to a group whose nature is defined by its relationship to a surrounding majority. Similarly, when we speak of "integration," we are describing the immersion of one group in a context where another group is prevalent. The traditional concept of "marginality" concerns the experience of the individual raised in one culture who finds himself in the midst of another. A group is not a "pariah people" as such but only in relation to the characteristics of the surrounding environment.

The aim of this chapter is to consider whether consonant or dissonant contexts have a bearing on the self-concept and, if so, why. Since not all environments are alike nor all self-concept dimensions the same, however, we shall consider four aspects or dimensions of the self-concept: (1) self-esteem, (2) instability of the self-concept, (3) group disidentification, and (4) salience.

## Dissonant Context and Self-Esteem

In order to examine these issues empirically, we shall draw upon data from two studies. The first—the study of Baltimore public school children from grades 3 through 12—was described in the previous chapter. The second body of data is based on a sample of 5,024 high school juniors and seniors from 10 high schools in New York State collected in 1960. These data do not purport to describe New York State high school students today but they serve to demonstrate the dissonance principle.

The sample was selected by stratifying the population of New York State public high schools by size of community, and choosing the sample of schools from this population through use of a table of random numbers. Three separate but overlapping questionnaires were administered alternately to the respondents; each student completed one questionnaire. Some of the New York State data, then, come from different questionnaire forms and are based on one-third of the sample.[3]

Both the Baltimore and the New York State data sets will be used to examine the effects on self-esteem of the following three contexts: social identity contexts, competence contexts, and values contexts. Unless otherwise identified, the tables in this chapter—in fact, throughout the book—are based on the Baltimore study. Tables based on the New York State study or other data sets are so identified in the table titles.

SOCIAL IDENTITY CONTEXTS

Since the individual in a dissonant context is, through circumstances usually beyond his control, sociologically deviant in his environment, this situation might have negative effects on his self-esteem. In examining this issue, we shall first consider three social identity contexts: religion, race, and social class.

*Religion.* The New York State adolescents were asked to recall the neighborhood in which they had lived longest when growing up and to indicate "the religious affiliation of most of the people in the neighborhood." Knowing the adolescent's own religious background, it was possible to determine whether he was reared in a predominantly dissonant, mixed, or predominantly consonant religious environment. Table 4–1 shows that those adolescents raised in dissonant religious environments were more likely to have low global self-esteem than were those raised in mixed or consonant religious environments.[4] Specifically, Catholics raised in non-Catholic neighborhoods had lower self-esteem than Catholics raised in Catholic or mixed neighborhoods; Jews raised in non-Jewish neighborhoods had lower self-esteem than Jews raised in Jewish or mixed neighborhoods; and Protestants raised in non-Protestant neighborhoods had lower self-esteem than Protestants raised in Protestant or mixed neighborhoods. The differences, although not large, were nevertheless consistent.

*Race.* One situation of dissonance or consonance, usually not recognized as such, is racial integration or segregation. The confusion is probably due to the evaluative connotations of the terms involved. We tend to think of dissonance as bad but integration as good, of consonance as good but segregation as bad. Yet technically—and, as

**TABLE 4-1.**

*Religious Dissonance and Self-Esteem (New York State)*

|  | Catholics | | Protestants | | Jews | |
|---|---|---|---|---|---|---|
|  | In non-Catholic neighbor-hoods | In Catholic or mixed neighbor-hoods | In non-Protestant neighbor-hoods | In Protestant or mixed neighbor-hoods | In non-Jewish neighbor-hoods | In Jewish or mixed neighbor-hoods |
| *Self Esteem* | | | | | | |
| Low | 41% | 29% | 31% | 25% | 29% | 18% |
| Medium | 30 | 25 | 27 | 30 | 10 | 23 |
| High | 30 | 46 | 42 | 45 | 61 | 60 |
| N = 100% | (37) | (458) | (164) | (241) | (41) | (80) |

we shall see, psychologically—the one is an exact expression of the other.

Table 4–2 compares black adolescents in Baltimore who attend segregated schools with those who attend desegregated schools. The results show that self-esteem is lower in the dissonant context.[5] At the junior high level, black children attending white schools had somewhat lower self-esteem than blacks attending predominantly black schools, and this difference was greater at the senior high level.[6]

The result is entirely consistent with the general pattern of findings

**TABLE 4-2.**

*Self-Esteem of Black Secondary School Students,*
*by Racial Composition of School*

|  | School racial context is predominantly . . . | | | | | |
|---|---|---|---|---|---|---|
|  | Black | White | Black | White | Black | White |
|  | Junior high schools | | Senior high schools | | All secondary schools | |
| *Self-Esteem* | | | | | | |
| Low | 19% | 23% | 12% * | 26% | 17% | 25% |
| Medium | 38 | 40 | 28 | 29 | 33 | 33 |
| High | 43 | 37 | 60 | 45 | 50 | 42 |
| N = 100% | (190) | (48) | (146) | (66) | (336) | (114) |

*See footnote 7, chapter 3, p. 98.

in the area. With regard to racial consonance or dissonance, St. John (1975) has reviewed 25 studies of desegregation and self-esteem. Nine showed negative effects, 4 positive, and 12 no effects or mixed effects. The negative effects of dissonance on self-esteem, however, are particularly likely to appear in studies using large, carefully selected samples. The very diversity of the self-esteem measures makes the conclusion all the more compelling. Powell and Fuller (1973) use the Total Net Positive Score of the Fitts Tennessee Self-Concept Scale, Bachman (1970) uses a 10-item Likert scale, and Beers (1973) uses the Coopersmith (1967) Self-Esteem Inventory. Even Crain and Weissman (1972), able to cite the longest list of advantageous outcomes associated with racial integration among black adults, point to one conspicuous exception—self-esteem.

*Social Class.* When we speak of higher, middle, or lower classes, we are of course using relative terms. These terms achieve meaning only by reference to certain anchorage points, as reflected in the story of an unfortunate Hollywood film writer in the fifties whose peers looked down on him because he earned "only" $50,000 a year. To some extent, certainly, one feels rich or poor by comparing oneself to those in the same environment. What effect, then, does socioeconomic consonance or dissonance have on self-esteem?

The evidence available is sparse and at best suggestive. Because of the familiar association of race and social class, Table 4–3 examines the effects of socioeconomic contexts among whites and blacks separately. Since Baltimore is largely a working-class city, respondents are divided into three social class groups, using the Hollingshead criteria: classes 1–3 (middle or upper middle), class 4 (working), and class 5 (lower). The schools these respondents attend are also classified according to their average socioeconomic levels. These *schools* have been divided into the "high" (mean socioeconomic status of 1.0–3.4) and the "low" (mean socioeconomic status of 3.5–5.0).

Considering first the white subjects, Table 4–3 shows that middle- or upper-middle-class children attending lower-class schools had significantly lower self-esteem than middle- or upper-middle-class children attending higher-class schools; that, in the working class, no difference appears; and that lower-class children in higher-class schools have lower self-esteem than lower-class children in lower-class schools. For both upper- and lower-class children, then, dissonant socioeconomic environments appear inimical to self-esteem. Among black subjects, the same

**TABLE 4-3.**

*Dissonant Socioeconomic Contexts and Self-Esteem, by Race*

|  | Whites | | | | | | Blacks | | | | | |
|---|---|---|---|---|---|---|---|---|---|---|---|---|
|  | Middle or upper middle class | | Working class | | Lower class | | Middle or upper middle class | | Working class | | Lower class | |
|  | School socioeconomic level | | | | | | School socioeconomic level | | | | | |
|  | Higher | Lower | Higher | Lower | Higher | Lower | Higher | Lower | Higher | Lower | Higher | Lower |
| *Self-Esteem* | | | | | | | | | | | | |
| Low | 26% * | 53% | 35% | 37% | 57% | 42% | 27% | 15% | 18% | 19% | 35% | 19% |
| Medium | 35 | 28 | 28 | 30 | 14 | 28 | 32 | 39 | 27 | 33 | 22 | 37 |
| High | 39 | 19 | 36 | 33 | 29 | 30 | 41 | 46 | 55 | 48 | 43 | 44 |
| N = 100% | (179) | (57) | (99) | (203) | (21) | (88) | (37) | (124) | (49) | (408) | (23) | (451) |

*See footnote 7, chapter 3, p. 98.

results obtain, with the exception of higher-class blacks in lower-class schools; their self-esteem is higher. Relative economic disprivilege, then, is associated with reduced self-esteem, but relative economic privilege shows an inconsistent association. Since we are aware of no other evidence on this question, the conclusions drawn must be tentative.

COMPETENCE CONTEXTS

Although it is customary in contextual analyses to treat sociodemographic variables (especially race and social class) as defining the context, there is ample theoretical (Durkheim, 1951) and empirical (Lazarsfeld and Thielens, 1958; Blau, 1960) basis for thinking that other variables shared within a given environment, such as attitudes, values, or competence levels, may also affect the group member. For example, how does the individual assess his various evaluated characteristics—how fast or slow, smart or dull, attractive or unattractive he is? Such judgments are not made in a vacuum; they are usually based on social comparisons, often with those in the immediate environment. This point can be illustrated by considering experience and abilities contexts.

*Experience Contexts.* In "The American Soldier" studies (Stouffer *et al.*, 1949), the Research Branch considered the following contexts: inexperienced troops in inexperienced outfits; equally inexperienced replacements in divisions otherwise composed of combat veterans; and veterans in divisions composed of other veterans. Among numerous questions put to these soldiers was one intended to assess the individual's confidence in his ability to take charge of a group in combat. As one might expect, the veterans were more likely to express confidence in their ability to take charge than were the inexperienced troops. The interesting comparison, however, was between the inexperienced troops in inexperienced outfits and the inexperienced troops in veteran outfits. It turns out that the inexperienced soldier in a dissonant context—that is, in a veteran outfit—was less confident that he could take charge of a group of men than was the inexperienced soldier in an inexperienced outfit. The competence was the same, but the contexts were different.

*Abilities Contexts.* It seems obvious that the individual's assessment of his abilities will depend in part on the abilities of those around him. Several examples of such effects were presented in the discussion of the social comparison principle in chapter 2. The studies described there suggested that whether or not a high school student aspires to attend college depends to some extent on his view of his academic ability; but his view of his academic ability, in turn, depends on how he compares with those around him (his class rank).

The most obvious competence context within a school system is the average level of marks. One might expect a student with a C average to judge himself more favorably if the school average were C+ than if it were A-. Empirically, however, it is surprisingly difficult to examine the significance of these contexts in relation to school marks for two reasons: first, there is not much difference in average marks from school to school and, second, very few students actually do have extreme averages (either A or D) whereas the great bulk of students—those with B or C averages—differ little from one another in self-esteem. We are thus obliged to confine our attention to the poor students—those with borderline or failing averages (D or F). Furthermore, because of insufficient whites in the Baltimore sample, we must confine our analysis to blacks in predominantly black schools. Is the self-esteem of such pupils affected by whether the average marks in their school are high or low? Table 4-4 suggests that it is. Borderline or failing pupils attending schools in which the average marks are relatively high are less likely to have high self-esteem than similar students in schools in which the average level of work is lower.

Since our data are based on the abilities context for only one small subgroup of the sample, they are little better than suggestive. The general principle, however—that the psychological impact of a given

**\*TABLE 4-4.**

*Self-Esteem of Poorly Performing Black
Pupils in Predominantly Black Schools
with Varying Average Marks*

|  | Black D or F pupils attending black schools in which average marks are relatively . . . | | |
|---|---|---|---|
|  | High | Medium | Low |
| *Self-Esteem* |  |  |  |
| Low | 33% | 28% | 27% |
| Medium | 54 | 39 | 25 |
| High | 13 | 33 | 48 |
| N = 100% | (24) | (36) | (48) |

\*See footnote 7, chapter 3, p. 98.

performance level depends on the general performance level in the environment—does appear to be a matter of common observation. For example, well-functioning retardates in retardate classes are reported to suffer disturbance when transferred to classes of normal children. Indeed, even high achievers may suffer rude blows to their self-esteem when finding themselves in the company of still higher achievers. This is an experience frequently reported by students who, having been outstanding high school seniors, suddenly find themselves to be mediocre as freshmen at highly select colleges.

VALUES CONTEXTS

Thus far we have treated the notions of dissonance and consonance as reflections of the relationship between a characteristic of the individual and the prevalence of that characteristic in his immediate environment. When we turn to the values context, the concept of dissonance is broadened to include not only the literal identity of the individual and the collective characteristics but also what might be called their "congeniality." Here, consonance or dissonance would be reflected in the relationship between a particular individual characteristic and the *value* attached to that characteristic in his environment.

An interesting example appears in Coleman's study (1961) of adolescents. Working with 10 high schools, Coleman first classified the schools into those in which athletic ability was highly valued and those in which it was not. Having sociometric data for the school population, he was also able to select the boys considered to be the best athletes in their schools. As a crude indicator of negative self-evaluation, he asked these boys whether they would "want to be somewhat different." Coleman found that, in the schools in which athletic achievement was highly valued, only 9 percent of the top athletes wanted to be somewhat different; in the schools in which athletic achievement was less highly valued, however, 15 percent of the top athletes wanted to be "somewhat different." It was thus not only the actual achievement of the individual—the boys in both groups were named as best athletes—but also the value attached to that particular ability in the context that influenced the feeling of self-dissatisfaction.

What is true of the value attached to traits is equally true of the value attached to statuses. Consider family structure. The most desirable family structure in American society is the intact nuclear family, the least desirable is the unmarried or abandoned mother and child. The normal expectation would be that children from the latter family structure would have lower self-esteem. Often overlooked, however, is the valuation or disvaluation of given statuses in different contexts.

In the Baltimore study, 30 percent of the black children came from separated or never-married families, compared to 6 percent of the whites. We do not know what proportion of these families represent illegitimacy, what proportion abandonment, and what proportion legal separation. Nevertheless, it seems reasonable to infer (see Farley and Hermalin, 1971) that some of these children stemmed from stigmatized origins. The probable consequences for the self-esteem of these children appear obvious.

But these consequences, it turns out, depend on the racial context in which this family structure is embedded. Comparing black children from separated or never-married families with black children from other family structures, Table 4–5 shows that, in predominantly black schools, there is no difference in self-esteem levels. But if black children from separated or never-married families attending mixed or predominantly white schools are compared with other black children in these racially dissonant schools, then they are much more likely to have low self-esteem. Specifically, 42 percent of them have low self-esteem compared with only 19 percent of the other black children in the mixed or white schools (Table 4–5). (A study of 798 black and white sixth and seventh grade pupils in a midwestern city by Simmons, Brown, Bush, and Blyth, 1977, using the same global self-esteem measure, shows strikingly similar results). Thus the impact of family structure on the black adolescent's self-esteem depends on whether he is in a black or white school context. Such family backgrounds, to be sure, are deplored by blacks as well as by whites (Rainwater, 1966) but, being more common, are probably less severely condemned. It may further be noted that white children from separated or never-married families, almost all of whom attend white schools, also have significantly lower self-esteem than do white children from intact families. Furthermore, although white children from these types of broken families are much more likely than white children from intact families to report that they have been teased by other children because of their families, black children from separated or never-married families, most of whom attend black schools, are not more likely than blacks from intact families to report such teasing.

**TABLE 4-5.**

*Self-Esteem and Family Structure of Black Children*
*Within Different Racial Contexts*

| | Black children in . . . | | | |
| | Predominantly black schools | | Mixed or predominantly white schools | |
| | Family Structure | | | |
| | Separated or never-married | Other | Separated or never-married | Other |
|---|---|---|---|---|
| *Self-Esteem* | | | | |
| Low | 19% | 18% | 42%  * | 19% |
| Medium | 32 | 36 | 33 | 37 |
| High | 49 | 46 | 25 | 44 |
| N = 100% | (290) | (601) | (36) | (162) |

*See footnote 7, chapter 3, p. 98.

# The Experience of Contextual Dissonance

Various contexts, then—social identity contexts, competence contexts, values contexts—may all bear on the child's and adolescent's global self-esteem. Before turning to the issues of instability of the self-concept and group disidentification, it is relevant to pause and to reflect on the question: just what is the actual living experience of dissonance that underlies this reduced feeling of self-worth? In seeking to understand these findings—or, indeed, the dissonance experience generally—three aspects of this experience will be considered: the dissonant communications environment; the dissonant cultural environment; and the dissonant comparison reference group.

DISSONANT COMMUNICATIONS ENVIRONMENT

Every context is a communications environment—a place where certain things are said or remain unsaid, where characteristic points of view prevail, where tacit assumptions underlie explicit messages, where shared norms, ideas, and values hold sway. Attitudes toward an endless variety of ideas and objects, including the self, are influenced by such communications. Specifically, then, to what kinds of communications would we expect a child in a dissonant context to be exposed? In order to highlight the general point, let us think of a prototypical prejudiced society, one characterized by widespread intergroup hostility—between blacks and whites, Hindus and Moslems, Anglos and Mexican-Americans, Irish Protestants and Catholics. What kinds of communications are likely to prevail in different contexts?

In the dissonant context, first of all, the child is obviously more likely to be exposed directly to racial, religious, or ethnic slurs, epithets, or attacks. In a variety of ways, the message is often imparted that he is considered inferior by virtue of his group membership. In a segregated area, on the other hand, the same child would be relatively insulated from such expression of prejudice; his own group, after all, does not revile him for his group membership. A poor child in a rich neighborhood might also be looked down on in a way that would not occur in a poor neighborhood.

The research evidence, although modest, is consistent with this obvious point. In the New York State study, adolescent subjects were asked whether they had ever been ". . . teased, laughed at, or called names by other children because of . . . your religion?" Not only were Jewish children who had been raised in predominantly gentile neighborhoods more likely to report such experiences than

**TABLE 4-6.**

*Dissonant Context and Subjection to Religious Prejudice,*
*by Religious Affiliation (New York State)*

| "When you were a child, were you ever teased, left out of things, or called names by other children because of your . . . religion? | Catholics | | Protestants | | Jews | |
|---|---|---|---|---|---|---|
| | In non-Catholic neighbor-hoods | In Catholic or mixed neighbor-hoods | In non-Protestant neighbor-hoods | In Protestant or mixed neighbor-hoods | In non-Jewish neighbor-hoods | In Jewish or mixed neighbor-hoods |
| †Ever | 22% * | 5% | 22% * | 6% | 48% * | 26% |
| Never | 78 | 95 | 78 | 94 | 52 | 74 |
| N = 100% | (37) | (454) | (162) | (238) | (42) | (78) |

*See footnote 7, chapter 3, p. 98.
†"Ever" refers to those who answered "often," "sometimes," or "rarely."

were Jewish children raised in predominantly Jewish (or mixed) neighborhoods, but so were Catholics raised in non-Catholic neighborhoods and Protestants raised in non-Protestant neighborhoods. As the Protestant data show, even *high* religious status does not exempt one from the punitive effects of prejudice in a dissonant context (Table 4–6).

The effects of *racial* dissonance are similar. The Baltimore City study revealed that, among junior and senior high school pupils, the more racially dissonant the black child's environment (either his neighborhood or his school), the greater the likelihood that he had been "teased, laughed at, or left out of things because of your race" (Table 4–7). Plainly, one effect of desegregation is to enhance exposure to prejudiced communications.

**TABLE 4-7.**

*Racial Teasing of Blacks, by Racial Composition
of School and Neighborhood*

| | Neighborhood is predominantly | | | *School is predominantly | | |
|---|---|---|---|---|---|---|
| | Black | Mixed | White | Black | Mixed | White |
| *Teased about race* | | | | | | |
| †Ever | 36% | 39% | 46% | 34% | 43% | 51% |
| Never | 64 | 61 | 54 | 66 | 57 | 49 |
| N = 100% | (1043) | (84) | (37) | (1044) | (70) | (146) |

*See footnote 7, chapter 3, p. 98.
†"Ever" includes "often," "sometimes," and "not very often."

Not only is the child in a dissonant context more likely to learn how much *he* is disliked, but he is also more likely to gain enhanced awareness of how much his racial, religious, or ethnic *group* is disliked. (As we shall see in our discussion of minority status in chapter 6, black children are not fully aware of how "most Americans" rank their group.) Here we shall consider the effect of the racial context on this awareness. What Table 4–8 shows is that the dissonant context has the effect of forcing the unpalatable reality into fuller awareness. The data show that it is much easier for blacks attending segregated schools to believe that their race is viewed favorably by "most people in America" than it is for blacks attending desegregated schools.

**TABLE 4-8.**

*Black Secondary School Students' View of*
*How "Most People in America" Rank Blacks,*
*by Racial Composition of School*

| | Black pupils in Junior High School | | Black pupils in Senior High School | |
|---|---|---|---|---|
| | School is predominantly | | | |
| | Black | White | Black | White |
| Believe that "most people" rank blacks . . . | | | | |
| First | 26% | 16% | 17% * | 2% |
| Second | 26 | 29 | 9 | 10 |
| Third | 28 | 19 | 28 | 26 |
| Last | 20 | 36 | 46 | 62 |
| N = 100% | (151) | (31) | (130) | (50) |

*See footnote 7, chapter 3, p. 98.

The reason seems evident. The black child in a white school is exposed to communications making it abundantly clear just how his race is regarded by whites. In the consonant racial context, on the other hand, it is much easier for the black child to deny, ignore, or misperceive his group's standing in the outside world. The same would presumably be true of Puerto Rican, Mexican-American, American Indian or any other children in different communications environments.

The critical point is that *no one is ever exposed to the totality of a culture*: he is only exposed to it as it is selectively filtered through his particular experience. Think of a black child attending a predominantly black school, living in a black neighborhood, interacting with black friends and relatives—in short, living in a black world. Talking and listening chiefly to blacks, what is likely to be called to his attention are the distinctive sources of black merit. Thus, he is likely to hear much of his racial heroes—those blacks who have achieved distinction, manifested conspicuous talent, elicited admiration—and to take pride in them. The same would be true of an Italian child in an Italian neighborhood, a Polish child in a Polish neighborhood, and so on. Experience in this communications environment may thus have the effect of convincing group members that their group holds a higher rank in the total society than it actually does.

DISSONANT CULTURAL ENVIRONMENT

Cultural dissonance is a time-honored theme in sociology, perhaps best expressed in Park's famous (1928) concept of the "marginal man." Briefly expressed, the idea is that people who have been socialized in one culture or subculture find themselves in environments in which other group norms prevail. In these groups, both the qualities of others and of the self are implicated in defining the individual as *different*.

To sharpen the point, let us again take an ideal type example. Consider a Ceylonese or Korean or French child who, for reasons beyond his control, finds himself in an American school. We might call him a "marginal child." Even in the total absence of direct prejudice against his race or nationality—an unlikely event—such a child may well feel strange, different, peculiar. He does not belong, he does not fit, he is "out of it." Until recently a respected conformist, well integrated into his own group, he now finds himself a deviant— rejected, laughed at, or despised because of his strangeness, peculiarity, or ineptitude.

Superimposed on these disadvantages is the fact that those qualities which he has been at such pains to cultivate, often constituting major sources of pride to him, are frequently unappreciated in the new cultural context. His talent at soccer, his elegant pronunciation, his respect for teachers, his academic competence may gain little applause in his new environment, where his awkwardness at baseball, his "peculiar" pronunciation, his misspelling of the "simplest" words, his ignorance of the heroes of sports, the noted figures of television, and "familiar" historical figures may elicit laughter or ridicule. The implications for self-esteem are obvious.

Two points deserve emphasis. The first is that, from the self-esteem viewpoint, the effect of the dissonant cultural context may be even more insidious than that of direct prejudice. The reason is that if others attack us directly, it is possible to mobilize our psychological defenses—to conclude that the other is stupid, bigoted, and ignorant, to counterattack with our own epithets, and so on. If, on the other hand, we accept and internalize the general standards and values of the dissonant context, then we may come to despise *ourselves*—to believe that we are strange and different, that we are inept at the skills and talents valued in the new environment, that we are ignorant of the things that count. This kind of attack is particularly devastating, for it is an attack from within.

The second point is that the cultural dissonance principle may apply to the higher status individual as well as to the lower—to the Protestant, the Anglo-Saxon or Northern European, the well-to-do child. He, too, may be disdained because he speaks peculiarly, holds different religious ideas, is ignorant of ingroup jokes and allusions, and eschews pot, cigarettes, or liquor. In other words, the Protestant is as much a "fish out of water" in Catholic or Jewish neighborhoods as the other way around, just as the higher-class child in a lower-class area is as strange, peculiar, and inept as the converse. And, indeed, the limited evidence available suggests that these groups, when placed in dissonant contexts, do have lower self-esteem than those in consonant contexts. As indicated earlier, Protestants in non-Protestant neighborhoods were likely to suffer from somewhat lower self-esteem than those in Protestant neighborhoods. Similarly, we saw (Table 4–3) that not only are lower-class children in higher-status schools less likely to have high self-esteem than those in consonant schools, but that, at least among whites, higher-class pupils in lower-class schools are also less likely to have high self-esteem. (This latter finding was not repeated for blacks.)

In sum, it is characteristic of cultural groups that they tend to feel united on the basis of shared norms, values, attitudes, and perspectives. Ease of communication and a sense of solidarity spring directly from such similarity of thought and feelings. The likelihood that an individual will be accepted into the group is thus a function not only of whether he is socially defined as different by virtue of his group membership, but also of whether he actually is different. For example, a Jewish child may learn from his parents, relatives, or friends, that it is extremely important to be a good student in school. If he is raised in a Catholic neighborhood, where, according to the New York State data, less stress is placed upon this quality, then he may be scorned by his peers as a "grind," an "eager beaver," and an "apple polisher." At the same time, he may place little value on being "tough," or a "good fighter"; these qualities, more highly valued in the group by which he is surrounded, may give him the reputation of being a "sissy." Furthermore, if he comes to value the skills, traits, or characteristics held in high regard by the majority in the dissonant context, but has not been socialized to cultivate these traits, then he personally may count himself as inadequate and inferior in these respects.

The evidence in this regard is flimsy but it is at least suggestive. The data come from the New York State study, which found that

Catholics living in predominantly Protestant areas experienced fewer self-esteem problems than Catholics living in predominantly Jewish areas; that Protestants in Catholic areas were less disturbed than Protestants in Jewish areas; and that Jews in Protestant areas were less disturbed than Jews in Catholic areas. Since the number of cases involved was very small, the data could not be considered reliable. Nevertheless, the evidence suggested that Catholics were culturally more similar to Protestants than they were to Jews; that Protestants were more similar to Catholics than they were to Jews; and that Jews were more similar to Protestants than they were to Catholics. Although the data are indirect, they are consistent with the interpretation that those reared in a "more dissonant" religious context showed lower self-esteem than those reared in a "less dissonant" context.

We would not, of course, necessarily expect the empirical findings based on a study of New York State adolescents in 1960 to obtain among a different age group at a different time and place, but we would still expect dissonant cultural contexts to have generally damaging effects. If, as is frequently asserted, subcultural differences are discernible between blacks and whites (Guterman, 1972; Pettigrew 1964; Rainwater, 1966), then blacks in white environments would be more deviant or nonconformist relative to those around them than those in black environments; this factor might help to account for the finding of lower self-esteem in the dissonant racial contexts.

DISSONANT COMPARISON GROUPS

If the cultural dissonance effect is based on whether we are the same or different from those around us, the comparison effect is based on whether we are better or worse. To a substantial extent, individuals draw conclusions about how good they are by comparing themselves with others. However, this theory is less useful than it might be since it fails to specify *which* others are used for comparison.[7] When social scientists imply that black, Mexican-American or American Indian children have lower self-esteem than the white or Anglo majority because they are more economically poor or less successful academically, they are implying that these children are comparing themselves with some general social average essentially descriptive of the white or Anglo majority. But this is the remote objective perspective of the social scientist, not the phenomenal field of the child. Young people, we believe, are likely to compare themselves with those in the *immediate* context, that is, with those around them.

Several otherwise puzzling conclusions make sense when seen in this light. Consider the fact that black children, despite their low average socioeconomic level, do not have lower self-esteem than whites (Wylie, 1978). The explanation of this mystery is largely to be found in the socioeconomic context. In order to gain some purchase on this issue in the Baltimore study, we asked the pupils whether they knew any children who were poorer than they and whether they knew any who were richer. As expected, black children, being disproportionately poor, were less likely than whites to say that they knew children who were poorer than they. Interestingly, however, the black children were also less likely to say that they knew *richer* children. This is puzzling. Why should an economically poor group be *less* likely to say that they know others economically superior to themselves? The answer is surprising: they *are* less likely to.

Let us start with the poorest black children (Class V on the Hollingshead scale). In order to look at the socioeconomic environment of these children, a mean socioeconomic status rating for the schools they attend was computed. If we look at Table 4–9, we see that 46 percent of these Class V black children attend schools where the average SES is almost the same as their own and that 49 percent are in schools one level higher; only 5 percent are in schools two levels higher. In other words, poor black children are mostly in schools in which most of the other chidren are equally poor or slightly better off. Among Class V *white* children, however, none of them attend schools in which the average socioeconomic level is at their own level; they are all poorer than average. To be sure, four-fifths of them are in schools in which the average is just one step higher, but an additional one-fifth are in schools which are two or more steps higher. The lower-class white child is thus more likely than the lower-class black child to suffer by economic comparison.

Class IV children show a similar pattern. Among blacks, two-thirds are in schools in which their own class level is average, but one-fifth are in schools in which their class level is above average, and only one-tenth in which their class level is below. Among Class IV whites, on the other hand, none attend schools in which their level is higher than the school average and one-third are in schools in which they are below average. Conversely, among higher-class children, blacks again compare more favorably with their peers than whites, i.e., the

**TABLE 4-9.**

*Mean Social Class of School by Child's Social Class, by Race*

| | *Blacks | | | | *Whites | | | |
|---|---|---|---|---|---|---|---|---|
| | Social class of child | | | | | | | |
| | Higher | Middle | Working | Lower | Higher | Middle | Working | Lower |
| *Mean social class level of school* | | | | | | | | |
| †High | 33% | 20% | 10% | 5% | 84% | 70% | 33% | 19% |
| Medium | 58 | 66 | 65 | 49 | 16 | 30 | 67 | 81 |
| Low | 8 | 14 | 25 | 46 | – | – | – | – |
| N = 100% | (36) | (131) | (475) | (496) | (97) | (149) | (314) | (115) |

*See footnote 7, chapter 3, p. 98.

†Schools with mean Hollingshead scores of 3.0-3.4 are "high," 3.5-4.4 are "medium," and 4.5-5.0 are "low." No mean school social-class level was above 3.0.

higher-class black is more likely than the higher-class white to be economically superior to his schoolmates.

The upshot is that for the poor black child in this study, there is rarely a sizable coterie of well-to-do school mates within range of visibility to call to attention his own relative deprivation. Poor as he is, he is essentially equal to those around him. The poor white child, on the other hand, is almost certain to find representatives of higher classes in his school—children who enter his experience, constituting part of his comparison reference group and making him more keenly aware of just how poor he is.

Contextual analysis thus produces a topsy-turvy world. White children, who are richer, are comparatively poor; black children, who are poorer, are comparatively rich. This paradox is reality. Viewed in the abstract, it is logically absurd; viewed in contextual terms, it makes perfect sense.

Academic performance again illustrates the power of the context to affect self-assessment. Table 4–10 shows that, in Baltimore, black secondary school children attending predominantly white schools obtained somewhat better marks than those in predominantly black schools. (This difference, however, is not significant and may be a product of sampling accident.) Since academic success is generally associated with higher self-esteem, one would expect them to have higher self-esteem. But though their school marks were higher than

**TABLE 4-10.**

*School Marks of* Black *Secondary-School Pupils,*
*by Racial Context of School*

| | Junior High School | | Senior High School | |
|---|---|---|---|---|
| | Black pupils in predominantly . . . | | | |
| | Black schools | White schools | Black schools | White schools |
| Marks in school | | | | |
| A–B | 32% | 39% | 23% | 32% |
| C | 51 | 52 | 58 | 57 |
| D | 17 | 9 | 19 | 11 |
| N = 100% | (167) | (44) | (144) | (65) |

the marks of black children in segregated schools, their self-esteem was lower.

The explanation, we believe, is again to be found in the dissonant comparison reference group. If blacks in white schools were comparing their school marks with blacks in black schools, then their self-esteem should be higher. But if they were comparing themselves with the majority whites in their own schools, then it should be lower, for their performance is below par in this context, especially in senior high school. The latter appears to be the case (Table 4–11).[8] Self-assessment is as much dependent on the performance of the comparison reference group as on one's own performance; and the evidence is consistent in indicating that, at least among children, those in the immediate context constitute the comparison reference group.

Unlike dissonant communications environments and cultural contexts, whose effects appear consistently deleterious for self-esteem, in principle certain dissonant comparison reference groups might be advantageous for self-esteem. What happens, for example, if someone enters a dissonant context whose average socioeconomic or academic level is, instead of being higher, lower than his own? Will this environment *raise* his level of self-regard? Thus far we have detected no such effect. Perhaps the advantages of having some source of superiority might be outweighed by the disadvantages of cultural dissimilarity and exposure to prejudiced communications. Where the individual compares *unfavorably* with those in the dissonant context, however, there is little question that his self-esteem tends to suffer as a consequence.

**TABLE 4-11.**

*Children's School Marks in Predominantly White
Secondary Schools, by Race*

| | Predominantly white . . . | | | |
|---|---|---|---|---|
| | Junior High Schools | | Senior High Schools | |
| | Black pupils | White pupils | Black pupils | White pupils |
| *Marks in school* | | | | |
| A–B | 39% | 43% | 32% * | 60% |
| C | 52 | 52 | 57 | 35 |
| D | 9 | 6 | 11 | 5 |
| N = 100% | (44) | (202) | (65) | (117) |

*See footnote 7, chapter 3, p. 98.

Although low self-esteem may be the major self-concept consequence of contextual dissonance, it is not the only one. Hence, three
others—self-concept instability, group disidentification, and self-concept salience—will also be considered, although more briefly

## Dissonant Context and Self-Concept Stability

In Chapter 2, we showed that instability of the self-concept was
significantly associated with certain measures of psychological disturbance even when self-esteem was controlled. Unfortunately, we
are aware of no research dealing with the *social conditions* likely to
generate instability. Here we shall deal with one of these conditions,
namely, the dissonant context.

RACE

The Baltimore study utilized a seven-item Guttman scale of
"stability of self-concept" with a Reproducibility Coefficient of
89 percent and a Scalability Coefficient of 65 percent (Sample items:
"How sure are you that you know what kind of person you really
are?" "A kid told me 'Some days I like the way I am. Some days I
do not like the way I am.' Do your feelings change like this?").[9]
Table 4–12 presents the relationship of contextual dissonance to

## TABLE 4-12.

*Stability of Self-Concept of Black Children
Attending Racially Desegregated or Segregated Schools,
by School Level (Self-Esteem Standardized)*

|  | Elementary | | Junior High | | Senior High | |
|---|---|---|---|---|---|---|
|  | School context is predominantly . . . | | | | | |
|  | Black | White | Black | White | Black | White |
| *Self-concept stability* | | | | | | |
| Unstable | 23% * | 43% | 26% * | 43% | 29% * | 43% |
| Medium | 57 | 36 | 54 | 33 | 53 | 40 |
| Stable | 20 | 21 | 20 | 24 | 18 | 17 |
| N = 100% | (603) | (25) | (145) | (43) | (181) | (65) |

*See footnote 7, chapter 3, p. 98.

stability, controlling on self-esteem by test factor standardization (Rosenberg, 1962b). At each school level, black children in dissonant racial contexts are conspicuously more likely to have unstable self-concepts than are black children in consonant contexts. The latter are not more likely to show high stability but to show medium stability.

### RELIGION

The New York State study lacked sufficient information on race, but data were available on the religious contexts of the neighborhoods in which the adolescents were raised (Table 4–13). Although the data do not show any association between contextual dissonance and instability [10] of self-concept among Jews (actually, there is an association, but it vanishes when self-esteem is controlled), there is some association between these variables among Protestants, and a stronger association among Catholics, even with self-esteem controlled. Religious dissonance thus appears to exercise some effect on self-concept stability, independent of self-esteem.

Our data do not permit us to determine precisely what it is about the dissonant context that is responsible for generating self-concept instability, but the point is of sufficient interest to warrant speculation. One possible reason, we suggest, is that *the dissonant context may fail to provide interpersonal confirmation for the individual's self-hypothesis.*

Although some self-attitudes are accepted without question, other self-attitudes require external confirmation. Among the various types

**TABLE 4-13.**

*Religious Context and Stability of Self-Concept
(Self-Esteem Standardized) (New York State)*

| | Catholics | | Protestants | | Jews | |
|---|---|---|---|---|---|---|
| | In non-Catholic neighborhoods | In Catholic or mixed neighborhoods | In non-Protestant neighborhoods | In Protestant or mixed neighborhoods | In non-Jewish neighborhoods | In Jewish or mixed neighborhoods |
| *Self-concept stability* | | | | | | |
| Stable | 29% | 41% | 37% | 43% | 47% | 48% |
| Intermediate | 28 | 30 | 30 | 29 | 26 | 26 |
| Unstable | 43 | 29 | 34 | 28 | 27 | 26 |
| N = 100% | (42) | (420) | (161) | (227) | (40) | (76) |

of evidence confirming or disconfirming a self-hypothesis, probably the most important is interpersonal; the individual requires confirmation of his self-hypothesis from the behavior of others toward him. For example, if a person considers himself likeable, then others must act as if they like him; if he sees himself as intelligent, others must show some respect for his intelligence; and so on. In general, others must behave toward the individual in a fashion consistent with his own picture of what he is like if he is to maintain a stable self-concept.

With regard to the racial context, it is possible that the greater self-concept instability of black children in dissonant settings may be due to the fact that the white children and teachers in these settings do not provide sufficient confirmation for the black child's self-concept. Whereas he may see himself as physically attractive, mild in disposition, and honest, at least some proportion of the prejudiced white children may not appreciate his good looks or other virtues and may act as though he is threatening, aggressive, or dishonest. Such behavior would call into question the black child's picture of himself. But any kind of prejudice—prejudgment—including the excessively positive, may have this effect. The "liberal" white teacher, who publicly asserts how smart and cute and otherwise wonderful the black child is, whether or not the child actually is like this at all, may generate equal uncertainty and confusion in the child and thereby produce an unstable self-concept.

All this, of course, is speculative, and outstrips our data. The results do suggest, however, that dissonant racial and religious contexts are associated with somewhat greater instability of the self-concept.

## Dissonant Context and Group Disidentification

The various elements of social identity considered earlier constitute bases of social definition and of self-definition. Although society virtually obliges the individual to define himself in terms of these socially recognized categories, this says nothing about whether he identifies or disidentifies with them. (We shall have more to say on the issue of group pride, group self-hatred, or identity conflict in chapter 7. Here, we confine our attention to the effect of the dissonant context on group identification).

The Baltimore study contained six items that, in one way or another, appeared to reflect group identification. Although some of these indicators are more questionable than others, all generally appear to reflect the dimension under consideration.[11] These six items were combined to form a score of racial identification, and the pupils in consonant or dissonant racial contexts were compared. The results (Table 4-14) show that in junior high schools the context appears to have no effect on racial identification, but in the elementary and senior high schools, contextual dissonance is asso-

TABLE 4-14.

*Racial Identification of Black Children in Desegregated or Segregated Schools, by School Level*

| | Elementary | | Junior High | | Senior High | |
|---|---|---|---|---|---|---|
| | School context is predominantly . . . | | | | | |
| | Black | White | Black | White | Black | White |
| *Racial identification* | | | | | | |
| Strong | 29% * | 7% | 26% | 23% | 35% * | 12% |
| Medium | 13 | 34 | 13 | 16 | 11 | 15 |
| Weak | 58 | 59 | 61 | 61 | 54 | 73 |
| N = 100% | (696) | (29) | (156) | (51) | (194) | (68) |

*See footnote 7, chapter 3, p. 98.

ciated with somewhat weaker identification. It should be stressed that only in the rarest of instances do we find the kind of strong racial disidentification implied by Kardiner and Ovesey (1951), Erikson (1966), Lewin (1948), and others; there is little evidence of real "black self-hatred" or flat racial disidentification. The effect of the dissonant context, rather, appears to be to make racial identification slightly more equivocal or uncertain than would otherwise be the case, but not to produce strong group rejection or shame of oneself as a black.

One possible reason for the observed degree of disidentification rests in Kurt Lewin's concept of "group boundaries." Speaking of the eighteenth-century European Jewish Ghetto, he noted (1948: 149–50):

The strength and the character of the boundary of the Jewish group has changed a great deal in the course of history. In the period of the Ghetto, there were clear, strong boundaries between the Jewish groups and the other groups. The fact that the Jews then had to live in restricted territories or towns of the country, and in certain districts within the town, made the boundaries obvious and unquestionable for everybody.

Similarly, it is possible that the boundaries of the black child's world are defined by *his* ghetto (if he lives in one) and, though it may be a clear consequence of the distasteful reality of prejudice and discrimination, it may have the unanticipated consequence of binding the members closer together, and enhancing identification with one's group.

Furthermore, since the context is a communications environment, the minority child in the dissonant context is likely to hear more deprecatory and fewer laudatory things about his group and about himself as a group member. One can understand how such communications could contribute to reduced self-esteem, increased awareness of the low esteem in which one's group is generally held in the society, and some weakening of group pride and identification.

## Dissonant Context and Salience of Components

Since data on the relationship of contextual dissonance to salience are lacking, this topic will be mentioned only briefly. Other things being equal, we suggest that self-concept components will tend to be more salient, that is, more in the forefront of attention, in a dissonant than in a consonant environment. A Southerner in the North, or a Northerner in the South, is far more conscious of his regional origin than when in his home environment. Similarly, an American in Europe is much more conscious of being an American than in his own country. When McGuire *et al.* (1978) asked subjects to "Tell us about yourself," ethnic groups in dissonant contexts were more likely than other people to mention their ethnic identity.

Similarly, in his study of American Jewish students who came to Israel to study at the Hebrew University, Herman (1970b) found that the "students became acutely aware of the juxtaposition, new to them, of Americans and Israelis. The salience of their Americanism increased, 79 percent declaring that they had occasion to think of themselves as Americans either 'more often' or 'very much more often' in Israel than in the United States."

Salience, of course, is not the same as importance. Traits or social identity elements may be extremely important to us even though we give them little conscious thought. Nevertheless, if some self-concept component is consistently salient, it may well *become* important, as exemplified by the colonial administrator who, even after returning to the homeland, is far more aggressively patriotic than his compatriots who never left their native land.

## Summary and Discussion

Insofar as contextual dissonance damages self-esteem, it illustrates certain of the principles enunciated in chapter 2. The dissonant communications environment involves the principle of reflected appraisals. Although the child tends to see himself as he is seen by others, the views that others have of him depend on whether the environment is consonant or dissonant. But contextual dissonance primarily illustrates the social comparison principle. Such comparisons may be in terms of superiority-inferiority, or of similarity-difference. In other words, the individual's self-esteem may be damaged either because his abilities or achievements compare unfavorably with those around him (dissonant comparison reference group), or because his habits or interests are different from those of other people in the environment (cultural dissonance). In each case, it is the relationship of the individual's characteristics to those prevalent in his environment that accounts for the observed effect.

Several reasons for the observed self-concept effects have been suggested. First of all, the consonant context is a congenial communications environment. Within it, the individual is protected from the prejudice of the outside world. The racial, religious, and nationality slurs, epithets, jokes, and "put-downs" so abundant in the general society are largely absent in this insulated environment.

Second, the consonant context represents a familiar, comfortable environment, for it is the culture into which the individual has been socialized. What people do, how they speak, what they believe in, what they want—all these are comprehensible and congenial, or alien and discomfiting, depending on whether the context is consonant or dissonant. Since groups insist on adherence to their norms, making social acceptance conditional upon such general conformity, the individual in the consonant environment is likely to have a feeling of belongingness, the one in the dissonant context to feel strange, "out of it," somehow "wrong." These effects, it may be noted, would apply to those in the high as well as to the low status groups.

Third, the dissonant context may represent an infelicitous comparison reference group. People in general, and children particularly, tend to compare their evaluated characteristics, such as their social class or school marks, with those around them. Thus, someone who stands well in these respects in a consonant context may find himself ranking low in a dissonant context. In principle, of course, the dissonant context might also represent a felicitous comparison reference group, but our empirical data are mute in this respect.

The dissonant context may also represent an environment of inconsistent reflected appraisals—of "interpersonal incongruence," in the words of Backman and Secord (1962). In such an environment there may well be a disconcerting mismatch between the individual's taken-for-granted self-concept, representing his fundamental framework for dealing with his world, and the messages about himself returned to him by others. Self-inconsistency, in Lecky's (1945) terms, or a shaky or unstable concept, is one possible outcome of such an environment.

The generalizations enunciated in this chapter, it should be stressed, will not necessarily apply to *all* conditions that can appropriately be characterized as dissonant or consonant. These conditions may depend on general attitudes toward the group (if prejudice against a group declines, the impact of dissonance on self-concept will probably be reduced); on the social desirability or functional necessity of a group member (the effects of being a minority female in a majority male college are not likely to be negative, nor are the effects of being bright in a dull group or expert in an amateur group). The conditions under which contextual dissonance will have noxious self-concept effects still remain to be specified.

The concept of "minority group" also assumes a different meaning when seen within this framework. This concept implicitly uses the total society, or some large and dominant portion of it, as the relevant context or environment. Our focus, on the other hand, has been on the "effective interpersonal environment" of the individual. We believe that children, at least, see themselves not in relation to the characteristics predominant in the total society but chiefly in relation to those around them—their schools, classrooms, neighborhoods, towns. The average Mexican-American child will tend to be economically below the children of the Gold Coast, academically below the wealthy private school children of New England, culturally dissimilar to the children of central Nebraska. But these others do not enter his experience in any relevant way, and their effect on him is likely to be modest. What counts more are those in his effective interpersonal environment—the people around him who respect or despise his group, are richer or poorer than he, do or do not share the culture into which he has been socialized. We believe that a number of erroneous inferences about the effects of minority group membership on children derive from the implicit use of the total society rather than the effective interpersonal environment as the relevant context.

## NOTES

1. This view, originally presented by Rosenberg (1962a), differs from that of Hessler, New, Kubish, Ellison, and Taylor (1971), who believe that the collective unit itself should be treated as dissonant or consonant.

2. Active debate persists regarding whether there is any such thing as a contextual effect (see Hauser, 1970; Barton, 1970; Sewell and Armer, 1966). Essentially, this

argument is based on the fact, frequently observed, that contextual factors appear to explain very little additional variance in the dependent variable after all individual factors are permitted to explain all the variance they can. Aside from whether this conclusion is a matter of statistical necessity or a reflection of social reality, it reflects far too narrow a conception of what "context" means. As empirically investigated, it refers exclusively to "climate" effects or "structural effects"—the direct influence of the environment on our attitudes independent of our own characteristics. As we shall see, there are contextual effects other than climate effects (see also Karweit, 1977; Campbell and Alexander, 1965) which merit investigation.

3. Although the schools were selected in terms of strict random criteria, the degree to which they represent the general population of New York State high school juniors and seniors at the time cannot be assessed. For one thing, parochial and private secondary schools are excluded. Second, students not present in school on the day of the questionnaire administration were excluded. The sample, then, refers only to juniors and seniors in public high schools present on the day of questionnaire administration.

4. This 10-item scale, which has a Coefficient of Reproducibility of 92 percent and a Coefficient of Scalability of 72 percent, has come to be known as the RSE (Rosenberg Self-Esteem scale) (see Wylie, 1974). A listing of the items and scoring procedure are presented in Appendix A-1. Sample items are: "I take a positive attitude toward myself" and "At times I think I am no good at all."

5. Some of the conclusions in this and subsequent chapters are predicated on the assumption that the Baltimore self-esteem scale has the same psychological meaning for both black and white children. A detailed discussion of this critical issue appears in Rosenberg and Simmons (1972: Ch. 2). The data presented there suggest that the scale taps the same psychological dimension and is equally reliable and valid for both races.

6. There is no difference among elementary school children, but the number of black elementary school children in predominantly white schools was so small that no conclusions were warranted. And, as one might anticipate, there were almost no white children in predominantly black schools.

7. Pettigrew (1967: 303) has concluded that "Considerably more empirical work is obviously needed to develop a systematic theory of social evaluation. Foremost among the unresolved issues is the central problem of referent selection."

8. Coleman *et al.* (1966) also found that black children in predominantly white schools performed *better* on standardized achievement tests than black children in black schools but that their "academic self-concept" scores were *lower.*

9. See Appendix A-4 for a description of the Baltimore self-concept stability items and scoring instructions.

10. The New York State self-concept stability scale is a five-item measure with a Coefficient of Reproducibility of 94 percent and Coefficient of Scalability of 77 percent. (Sample items: "Do you ever find that on one day you have one opinion of yourself and on another day you have a different opinion?" "I have noticed that my ideas about myself seem to change very quickly.") A description of the New York State stability items and scoring instructions appear in Appendix A-3.

11. These indicators of racial group identification are described in Appendix B-2.

# 5

## Social Class and Self-Esteem Among Children and Adults*

THE CHILD'S self-concept is affected not only by his face-to-face relationships with those directly surrounding him—the family members so likely to represent significant others, the schoolmates and neighborhood peers in his context—but also by the broader social structure, especially its system of social class stratification. If the essential characteristic of social class, in the Weberian sense of "status group" (Gerth and Mills, 1946), is unequal prestige, then one would expect those looked up to by society to develop a high level of self-respect, and vice-versa.

Yet we cannot adequately appreciate how a social structural variable like social class affects the child's self-concept without taking account of the fact that it may affect the adult's self-concept differently. For it is apparent that social class, to children, signifies a radically different set of social experiences and is endowed with entirely different psychological meaning than social class to adults. The most

* This chapter was written in collaboration with Leonard I. Pearlin.

obvious demonstration of this point is that, for the adult, social class is achieved (at least in principle), whereas for the child it is unequivocally ascribed. The fundamental meaning of social class thus differs for children and adults. The bulk of this chapter will be devoted to a consideration of how the four principles of self-concept formation discussed in chapter 2—social comparison, reflected appraisals, self-attribution, and psychological centrality— help to explain why social class should have a *different* effect on the self-esteem of children and of adults.

## Social Class and Self-Esteem: Children, Adolescents, Adults

In order to compare children and adults, this chapter draws on two bodies of data. The children's data are drawn from the Baltimore study. The adult information is based on scheduled interviews with a sample of 2,300 people aged 18–65 representative of the census-defined urbanized area of Chicago (U.S. Bureau of the Census, 1972), which includes sections of northwestern Indiana as well as some of the suburban areas of Chicago.[1] In the adult study, global self-esteem is measured by the 10-item self-esteem measure used in the New York State study, scored according to the Likert method. Table 5–1 compares the relationships between socioeconomic status and global self-esteem among the Baltimore school pupils and the Chicago adults. Because race operates as a suppressor variable[2] (Rosenberg, 1968, 1973b; Davis, 1971) in the school pupil data, its effect is statistically controlled for both children and adults in this table.

Table 5–1 shows virtually no association between social class, measured by the Hollingshead "Index of Social Position" (Hollingshead and Redlich, 1958: 387–397), and self-esteem among the 8–11 year olds (partial gamma = +.029) and a modest association for early adolescents (for 12–14 year olds, partial gamma = +.072) and for later adolescents (for those 15 or older, partial gamma = +.102). In the adult data, on the other hand, the relationships are considerably larger: controlling for race, the relationship (partial gamma) of education and self-esteem is +.197, of occupation and self-esteem, +.160, and of income and self-esteem, +.233. In order to determine whether the associations in these various groups are significantly different from one another, Pearson correlations have been compared (Blalock, 1972: 405–407). For the 8–11 group, the correlation is significantly different from all three adult correlations at the .01 level; for the 12–14 group, the differences are all significant at the .08 level; and for the 15+ group, the differences are all significant at the .05 level. The three children's age groups, however, are not significantly different from one another.

**TABLE 5-1.**

*Social Status and Self-Esteem Among Baltimore School Pupils
and Chicago Adults (Controlled on Race)*

| | Baltimore School Pupils | | | Chicago Adults | | |
|---|---|---|---|---|---|---|
| | Social Class and Self-Esteem | | | Education and self-esteem | Occupation and self-esteem | Income and self-esteem |
| | Age | | | | | |
| | 8-11 | 12-14 | 15+ | | | |
| Partial gamma | +.029 | +.072 | +.102 | +.197 | +.160 | +.233 |
| Number | (699) | (571) | (468) | (2288) | (1689) | (1988) |

Results of earlier research have been too inconsistent to permit simple comparisons with our findings. Among younger children, several studies show *inverse* relationships between social class and self-esteem (Trowbridge, 1970, 1972; Piers, 1969; Soares and Soares, 1969), and some show null or close to null relationships (Coleman, et al., 1966; St. John, 1971). Among adolescents, Bachman (1970), Rosenberg (1965), and Jensen (1972) show small associations between social class and self-esteem; Epps (1969) shows null or close to null relationships; and Soares and Soares (1972) show an inverse relationship. Finally, among adults, Weidman, Phelan, and Sullivan (1972) and Yancey, Rigsby, and McCarthy (1972) report clearly positive relationships, while Luck and Heiss (1972) find positive associations within educational groups (though not across the total SES range), and Kaplan (1971) finds no overall association (although relationships do appear under specified conditions).

To these findings must be added the results of two unpublished studies of adults. One of these is Middleton's (1977) investigation of over 900 black and white adult men, which showed the following zero-order correlations of social class with self-esteem: education, r = .359; occupational status, r = .366; and income, r = .378. Although these relationships decline somewhat when controls are introduced, they remain reasonably strong—certainly stronger then in any of the children's data.

The second study is Kohn and Schooler's (Kohn, 1969; Kohn and Schooler, 1969) nationwide sample of 3,101 working men. Socioeconomic status was measured by the Hollingshead scale and self-esteem by a Guttman scale of global self-esteem. The correlation, which is significant at the .01 level, was r = .1929.[3]

Since adult and children's studies often use different measures of self-esteem, different measures of social class, different types of samples, different measures of association, and different statistical controls, one cannot compare the exact sizes of the relationships. Nevertheless, we believe the available evidence is reasonably consistent with the Baltimore and Chicago findings showing virtually no association between social class and self-esteem among pre-adolescents, a modest relationship among adolescents, and a moderate relationship among adults.

The most compelling explanation of this relationship among adults is that advanced by Kohn (1969) and Kohn and Schooler (1969). Their study demonstrated that members of different social classes differ radically in the types of occupational activities in which they are actually engaged. The job imperatives are such that the work of those in higher status positions is characterized by a high level of occupational self-direction—an opportunity to make one's own decisions, to exercise independent judgment, to be exempt from close supervision—in large part because of the substantive complexity of the work. Kohn (1969: 184) showed that when occupational self-direction was controlled, the originally significant relationships of social class to two self-esteem factors vanished almost completely.[4] From these findings, we may deduce that one reason social class has little effect on the self-esteem of children is that children are not yet exposed to the class-related occupational conditions that shape self-esteem.

This chapter presents additional theoretical reasons for expecting the observed variations across the age span, exemplified in Table 5–1, to occur. Two general explanations will be suggested: the first is that social class organizes the interpersonal experiences of children and adults in different ways, the second that social class is interpreted within different meaning frameworks by children and adults. These two factors, we believe, can help us to understand why *the identical social structural variable (social class) can produce different psychological outcomes (self-esteem) for different social groups (children and adults).*

This is not to suggest that different general principles of self-esteem formation operate among children and adults. On the contrary, the four basic principles of self-concept formation used throughout this work—reflected appraisals, social comparisons, self-attributions, and psychological centrality—apply to children and adults equally. But as these principles actually enter the lives of children and adults, they explain why social class has *different* effects on the self-esteem of children and adults.

## Social Comparison Processes

When one thinks of social class (in Warner's sense of social prestige), one simultaneously must think of social comparison, for the very meaning of social class is founded on the idea of relative position—of superior or inferior, higher or lower, better or worse. The special relevance of social class for self-esteem obviously rests in the comparison of one's prestige with that of other people. Concretely, however, with whom does the individual actually compare his social status? Does he compare himself with some broader societal average or with the standards of some narrower subgroup? In order to examine this question, we must again consider the nature of the *immediate social context*. In particular, the socioeconomic comparison will depend on the *homogeneity* or *heterogeneity* of the socioeconomic environment in which the individual is ensconced. Under conditions of visible social class heterogeneity, the higher classes may pride themselves on their superior status while the lower classes may be made painfully aware of their relative inferiority (Kaplan, 1971: 46). In the company of peers, on the other hand, status neither enhances nor diminishes self-esteem. In other words, in a completely classless—or single-class—environment, class will have no effect on self-esteem.

One reason social class makes less difference for the self-esteem of the child, we suggest, is that his interpersonal environment is more socioeconomically homogeneous. In examining this proposition, two questions will be considered. First, is the young child more likely than his seniors to *perceive* his environment as socioeconomically homogeneous? Second, is the socioeconomic environment of the young child *actually* more homogeneous?

The Baltimore data provide rather clear evidence that children tend to perceive their socioeconomic environments as homogeneous. Asked "Would you say that most of the kids in your class at school are richer than you, the same as you, or poorer than you?" fully 93 percent of the subjects said "the same." Only a miniscule proportion of subjects of any class considered themselves either above or below most of those in their school environments.

Furthermore, subjects were asked: "Do you know any kids whose families are richer than your family?" and "Do you know any kids whose families are poorer than your family?" "Any" is a large umbrella, indeed. It turns out that the younger the child, the less likely is he to say that he knows *either* richer children *or* poorer children. Among the youngest, 24 percent said that they knew richer kids and 32 percent that they knew poorer kids. Among the oldest, 66 percent said that they knew richer kids and 71 percent that they knew poorer kids. As the child grows older, he apparently becomes increasingly conscious of economic inequality.

Although the relative obliviousness of younger children to economic inequality is due in part to limited social learning, the interesting point is that it also reflects reality. Children are more likely to perceive their environments as socioeconomically homogeneous because these environments are in reality more homogeneous. Two items of evidence point to this conclusion.

First, most children, of whatever SES level, attend schools whose average socioeconomic level is similar to their own. Irrespective of age, the great bulk of lower-class children attend schools whose average socioeconomic status is low, higher-class children where it is high. Among the younger children, only 3 percent of the lower-class (Hollingshead Class V) children compared with 50 percent of the higher-class children (Hollingshead Class I and II) attend higher class (school SES mean 3.0 or above) schools; in the 12–14 age group, the corresponding figures are 5 percent and 53 percent; and among those 15 or older, the figures are 13 percent and 82 percent. Expressed another way, the relationships between the individual's class and the social class mean of his school for the various age groups is as follows: 8–11 years, gamma = .6173; 12–14 years, gamma = .6395; 15+ years, gamma = .5891.

When children of whatever age say that their SES is pretty much "the same" as those in their environments, then, they are simply describing reality. But what about the fact that adolescents are more likely to be aware of *some* socioeconomic heterogeneity in their environments—to say that they know at least some children who are richer or poorer than themselves? Is greater socioeconomic hetero-geneity more likely to be a part of their experience? If we look at the SES standard deviations of schools attended by younger and older children, we see that these tend to be larger for older subjects (Table 5–2). For example, whereas 47 percent of the youngest children attended schools with a relatively small standard deviation (under 0.7), this was true of only 20 percent of the oldest subjects. This is not surprising. High schools are not only much larger than elementary schools but also, drawing from a much wider geographi-cal area, tend to recruit children of more diverse socioeconomic backgrounds. Hence, it is understandable that the older child should *actually* be more likely to encounter a wider range of socioeconomic statuses in his school.

The younger the child, then, the greater the perceived and actual homogeneity of his school (and probably neighborhood) environ-ment. If children's interpersonal environments are such that most SES comparisons are with equals (or perceived equals), it is under-standable that, feeling neither above nor below others, their social class should neither raise nor lower their self-esteem.

The adult experience, to be sure, is not entirely dissimilar, since a good deal of adult interaction also goes on within socioeconomically

**\*TABLE 5-2.**

*Age by Standard Deviation of School SES*

|  | Age | | |
| --- | --- | --- | --- |
|  | 8-11 | 12-14 | 15+ |
| *School SES standard deviation* | | | |
| †Small | 47% | 35% | 20% |
| Medium | 46 | 61 | 51 |
| Large | 7 | 4 | 29 |
| N = 100% | (821) | (649) | (514) |

\*See footnote 7, chapter 3, p. 98.
†Small = < 0.7; medium = 0.7-1.1; large = > 1.1.

homogeneous environments. Yet it is clear that adults are more conscious of, and actually more exposed to, socioeconomic inequality.

It will be recalled that in the Baltimore study, 93 percent of the children, asked whether they were richer, poorer, or the same as most schoolmates, replied "the same." In the Chicago adult study, subjects were asked: "Would you say your total family income is higher, lower, or about the same as the following groups: Most of your friends; most of your relatives; most of your neighbors?" As one might expect, many adult subjects also considered themselves average in terms of these comparable membership groups. Nevertheless, 43 percent saw themselves as either richer or poorer than most of their friends, 61 percent thought they were above or below most of their relatives, and 49 percent considered themselves economically superior or inferior to most of their neighbors.

Furthermore, it is this feeling of economic superiority or inferiority that partly accounts for the relationship between social class and self-esteem among adults. Table 5–3 shows that the zero order gamma of income and self-esteem is .2485; controlling on whether they consider themselves richer or poorer than friends,[5] the partial gamma is .1509. For relatives, the corresponding figures are .2594 and .1834; for neighbors, .2462 and .1598. In other words, among those adults who consider their income unequal to that of their "peers," between 29 percent and 39 percent of the relationship between income and self-esteem is apparently attributable to social comparison processes.

TABLE 5-3.

*Income and Self-Esteem, Controlling on Comparison*
*with Peers, Among Those Perceiving Their Income as*
*Higher or Lower Than Others (Chicago)*

|  |  | Income and Self-Esteem | | |
| --- | --- | --- | --- | --- |
|  | N | Zero-order gamma | Partial gamma | Percent reduction |
| Controlling on whether subject considers self richer or poorer than ... |  |  |  |  |
| Friends | (421) | *.2485 | .1509 | 39% |
| Relatives | (594) | *.2594 | .1834 | 29% |
| Neighbors | (423) | *.2462 | .1598 | 35% |

*See footnote 7, chapter 3, p. 98.

These data, however, are an inadequate reflection of the degree of socioeconomic heterogeneity that actually characterizes the adult environment. The reason is that it is not among friends, neighbors, and relatives, after all, that the adult experiences the greatest SES heterogeneity (by and large, these are his peers) but rather in the world of work. Within the institutional structures of business, industry, and government, life is organized in hierarchical fashion. There is virtually no way in such environments to evade the hard reality of differences in power, prestige, and possession; everyone knows who gives the orders and who takes them, who commands great respect and who little, and who is paid well, who poorly. The actual experience of an adult in the world of work thus inevitably calls to his attention his place in a recognized stratification system. Is it any wonder that the adult's social class should show a stronger relationship to his self-esteem than the child's?

The general principle of social comparison, of course, applies equally to children and adults. The self-esteem of both groups is influenced by comparing themselves with those around them. But, given their different social roles, the *socioeconomic* environments of children and adults differ radically. Schooling is, after all, the main business of the child's life, working the main business of the adult's. In the school, children (especially younger children), whatever their own SES levels, tend to be more or less socioeconomically equal to most of those around them; there is nothing in this to raise or lower their self-esteem. In the world of work, on the other hand, the social comparisons adults make place them above or below others and may understandably affect their feelings of self-worth. Both children and adults make social comparisons, but the differing structures of interpersonal relations in school and workplace makes SES far more relevant to the adult's than to the child's self-esteem.

## Reflected Appraisals

If people tend to see themselves through the eyes of others, it would appear to follow that those standing high in the status hierarchy, commanding the respect of their fellows, should in the long run develop positive attitudes toward themselves; the converse would be true, of course, of those with low prestige in the society. This question is, however: do those people with whom we predominantly interact, and about whose opinions we most care, judge us by our *social class position?* Do they respect us because we are in a higher class, lack respect for us because we are in a lower class?

In order to deal with this question in a general sense, we must ask which categories of people characteristically enter the individual's experience by virtue of his location in the social structure. *Who* has opinions of us and *whose* opinions we care about is not so much a matter of accident as of our "role-sets." As we pointed out earlier, the role-set of the child in American society chiefly involves patterned relationships with his mother, father, siblings, other relatives, teachers, friends, and classmates.

The relevance of these observations is this: whereas the adult's role-set consists largely of status unequals, the *various individuals who constitute the child's role-set are almost all of his own socioeconomic status.* This is by social definition true of his mother, father, and siblings; probably true of his school and neighborhood peers; and possibly true of his teachers and more remote relatives. This is not to say that these others do not judge and evaluate the child. On the contrary, they may rank him high in looks, medium in intelligence, low in cleanliness; may admire his athletic ability, have a poor opinion of his neatness; may approve of his friendliness, deplore his laziness. But one thing that parents, siblings, and most peers neither admire nor deplore, approve nor disapprove, respect nor disrespect is his social class. No father looks up to his child as a member of the social elite, no brother looks down on him because of his working class origins. Yet mothers, fathers, siblings, and peers are precisely those people whose opinions count most to him, and who, as research indicates, most powerfully affect his self-esteem. Among possible status unequals, teachers alone are likely to have comparable influence. In sum, if significant others neither look up to nor down on the child by *virtue of his socioeconomic status,* then, adopting the viewpoint of these others, his objective SES should have little impact upon his feeling of self-worth.

Contrast this situation with the adult status-set, particularly his occupational role-set. To return to Merton's (1957) example of the teacher, he or she may be treated deferentially by the illiterate parent, imperiously by the highly educated parent; may be condescended to by the school principal, addressed respectfully by the janitorial staff. Admittedly, we may not greatly care about or respect the views of some of these people, but neither are their attitudes entirely irrelevant. Since work is such an important part of the adult's life, what those who enter his work experience think of him is likely to have a bearing on his self-attitudes.

Furthermore, consider the adult's family statuses. A father cannot consider his child an economic failure, but a child can consider his father to be one. A wife may express satisfaction with her husband's occupational success or dissatisfaction with his failure, and so, in fact, may the adult's own father, mother, brothers, sisters, and other relatives. Unlike the child, many of the individuals who enter the adult's life by virtue of his structural location in the society—his various role-sets—do view him in social status terms.

In addition, the adult's status-set is obviously much broader than that of the child. For the child, most of his interactions with others are exhausted by the categories of others that enter his role-set as child; for the adult, role-sets beyond occupation and family are involved. Hence, the adult's status-set brings him into contact with many more people who view him in status terms; to the extent that he sees himself from their viewpoints, his social status should affect his self-esteem.

Again we see that the identical principle—reflected appraisals—helps explain why social class should have *different* effects on the self-esteem of children and adults. The child's self-attitudes are very strongly influenced by others' attitudes toward him—every bit as strongly as the adult's—but if others do not judge the child in status terms, then status will have little self-esteem effect.

As we pointed out in chapter 2, the principles of social comparison and reflected appraisals are distinctively social, referring to what other people think of us or to how we compare ourselves with them. But the significance of social class for the feeling of self-worth also depends on the subjective meaning assigned to the objective fact of social class and on the position of social class in the individual's self-concept structure. The principles of self-attribution and of psychological centrality aptly illustrate this point.

## Self-Attribution

Like the principles cited above, self-attribution would also lead us to expect higher self-esteem among the socioeconomically successful. As noted earlier, this principle holds that people assess themselves by observing their behavior and its outcomes. A person who ranks high in the status hierarchy would be expected to have high self-esteem not only because he compares favorably with those around him or because he commands their admiration and respect, but because he himself interprets his success—whether based on the accumulation of money, prestige, or power—as evidence of how good he is. When William James (1890: 306–7) asserted that in the long run a man's self-esteem will come to correspond to his actual success or failure in life, he meant at least in part that this was based on the individual's own judgment of his personal accomplishments.

But this represents the assessment of *his* accomplishments, not the accomplishments of his ego-extensions (except insofar as the accomplishments of others are interpreted as the outcome of one's own efforts). But for the adult, we have pointed out, socioeconomic status is achieved, for the child, ascribed. For the former this status (high or low) is *earned*—the outcome of his efforts and his actions; for the latter, it is *conferred*—the product of another's accomplishments. If self-attribution theory suggests that people evaluate themselves largely in terms of their *own* behavior or its outcomes, it is certainly easy to understand why social class should be more closely tied to the self-esteem of adults than of children.

Children's self-esteem, we believe, is just as dependent on achievement as adult self-esteem; but this achievement is their own, not that of their parents. Unfortunately, it is not easy to test this proposition, because there are so few areas of recognized achievement available to the child. There are, however, two visible outcomes of his behavior which the child can use (as do other people) for drawing conclusions about certain aspects of his own worth. The first is school marks; the child may use this overt evidence to draw conclusions about how smart, or at least how good a student, he is. The second is election as an officer in a school club or organization; the child will certainly take account of this success in judging how popular he is or whether he has qualities of leadership.

As far as academic success is concerned, the relationship of school marks and academic self-concept is a consistently strong one; the studies cited in chapter 2 showed correlations varying between .48 and .57. It is consistent with self-attribution theory to think that the child has drawn these conclusions about himself in considerable part by observing his school marks, just as he would draw similar conclusions about other people.

Since elementary schools rarely have clubs or organizations, and since many high school students do not join such groups, the relationship between such election to a club office and self-esteem can be assessed only for the limited number of high school students who are club members. The New York State study shows that three-fifths of the highest self-esteem respondents had held some elected post in a club or school organization, compared with slightly over one-third of those with the lowest self-esteem. With one minor exception, each step down the self-esteem scale is matched by a decreasing proportion of group officers (Table 5–4). Essentially, the same results appear in the Baltimore study although, in that context, the number of club members is so small that the association is not statistically significant.[6]

**\*TABLE 5-4.**

*Self-Esteem and Election to Officerial Post in Club or School Organization (New York State)*

|  | Self-Esteem | | | | | | |
|---|---|---|---|---|---|---|---|
|  | High 0 | 1 | 2 | 3 | 4 | 5 | Low 6 |
| Held elected position | 60% | 59% | 52% | 49% | 49% | 42% | 36% |
| Never held elected position | 40 | 41 | 48 | 51 | 51 | 58 | 64 |
| N = 100% | (247) | (385) | (343) | (223) | (111) | (53) | (22) |

\*See footnote 7, chapter 3, p. 98.

One reason social class has so little bearing on the self-esteem of the child, then, is that the societal stratification system does not represent an arena of personal accomplishment. What the child himself has wrought appears to bear on his self-esteem; his father's achievements, on the other hand, are not his own.

But the identical principle applies to adults. What the adult has achieved may affect his feeling of self-worth but what *his* father has achieved is largely irrelevant. For example, the Chicago study showed a relationship of gamma = .18 between the adult's own occupation and his own self-esteem and of gamma = .07 between his father's occupation and his own self-esteem. The point is that the partial gamma of father's occupation and own self-esteem, *controlling on own occupation*, is only .01. Even if father's status helped one to get ahead (or otherwise) in the first place, it is the proximal fact of one's *own* success or failure, not the distal fact of paternal prestige, that directly affects the adult's self-esteem.

Like the other principles of self-esteem formation, self-attribution theory applies equally to children and adults. The reason social class affects the self-esteem of adults more than of children is that different interpretations are assigned the same structural facts by members of these age categories. The general implication is that the impact of a structural arrangement on a psychological disposition depends on the meaning of that objective fact to the individual involved.

## Psychological Centrality

Since, as indicated in the first two chapters, the various parts, elements, or components of the self-concept are of unequal importance to the individual, an understanding of the relationship of social class to self-esteem must take account of the location of social class in the individual's self-concept. Like any self-concept component—whether a trait (sympathetic or tactful), a social identity element (race or religion), a label or type (drug addict or intellectual)—the meaning and significance of social class depends on how central or peripheral it is to the individual.

For example, in Table 5–1, we noted that adults with higher income are more likely to have high self-esteem than the economically less successful. But the strength of this relationship, it turns out, depends on how much importance the individual attaches to money. To determine the degree of this importance, the Chicago adults were asked to agree or disagree with the statement: "One of the most important things about a person is the amount of money he has." Table 5–5 shows that the more strongly the individual agrees with this statement, the more powerful is the relationship between his actual income and his self-esteem. Among those who strongly agree, the relationship of income to self-esteem is gamma = .52; for those who agree or disagree somewhat, the gammas are .39 and .37, respectively; for those who strongly disagree, gamma = .21.

Table 5–1 also shows a clear though moderate association between adult occupational status and self-esteem. But this association, we find, likewise depends on the importance attached to status. Subjects were asked: "How important is it to you to move to a higher

## TABLE 5-5.

Self-Esteem, Level of Income, and Importance Attached to Money (Chicago)

| | Consider Money "One of the Most Important Things" | | | | | | | | | | | |
| | Strongly Agree | | | Somewhat Agree | | | Somewhat Disagree | | | Strongly Disagree | | |
| | Family Income (in Thousands) | | | | | | | | | | | |
| | 16+ | 8-16 | 8– | 16+ | 8-16 | 8– | 16+ | 8-16 | 8– | 16+ | 8-16 | 8– |
|---|---|---|---|---|---|---|---|---|---|---|---|---|
| *Self-Esteem* | | | | | | | | | | | | |
| High | 72% | 26% | 16% | 35% | 24% | 17% | 42% | 23% | 15% | 51% | 47% | 32% |
| Medium | 6 | 40 | 22 | 38 | 24 | 13 | 30 | 22 | 18 | 24 | 22 | 24 |
| Low | 22 | 35 | 62 | 27 | 51 | 70 | 28 | 55 | 67 | 25 | 31 | 43 |
| N = 100% | (18) | (43) | (69) | (48) | (115) | (83) | (81) | (224) | (123) | (325) | (563) | (292) |
| | *Gamma = .5222 | | | *Gamma = .3868 | | | *Gamma = .3688 | | | *Gamma = .2140 | | |

*See footnote 7, chapter 3, p. 98.

## TABLE 5-6.

*Self-Esteem, Occupational Status, and the Importance of Status Aggrandizement (Chicago)*

| | Importance of Moving to Higher Prestige Class | | | | | | | | | | | |
|---|---|---|---|---|---|---|---|---|---|---|---|---|
| | Very or Somewhat Important | | | | Little Importance | | | | No Importance | | | |
| Self-Esteem | Professionals | Clerks, technicians, sales | Skilled | Un- and semi-skilled | Professionals | Clerks, technicians, sales | Skilled | Un- and semi-skilled | Professionals | Clerks, technicians, sales | Skilled | Un- and semi-skilled |
| High | 43% | 34% | 24% | 21% | 45% | 38% | 36% | 29% | 53% | 42% | 49% | 38% |
| Medium | 25 | 26 | 32 | 19 | 24 | 25 | 23 | 26 | 22 | 29 | 21 | 20 |
| Low | 32 | 40 | 44 | 60 | 31 | 37 | 41 | 45 | 25 | 29 | 29 | 42 |
| N = 100% | (47) | (129) | (82) | (99) | (97) | (191) | (152) | (144) | (108) | (197) | (160) | (123) |
| | *Gamma = .2498 | | | | Gamma = .1413 | | | | *Gamma = .1074 | | | |

*See footnote 7, chapter 3, p. 98.

prestige class?" Among those who considered it very or somewhat important, the association is gamma = .25; but among those considering it unimportant, gamma = .11 (Table 5–6).

The impact of social class or status on global self-esteem, then, depends in part on its psychological centrality for the individual.[7] The relevance of this point in the present context is that social status affects adults' more than children's self-esteem simply because *social status is more psychologically central to the adult than to the child.* Adults are more aware of, attuned to, concerned with social status than children; the adult is more likely to perceive his world in stratification terms, to be alert and sensitive to his own and others' social rank. Several different items of evidence support this point.

One indication of whether class plays a major role in the individual's cognitive structure is simple familiarity with the vocabulary of stratification. Although it is possible to be class conscious without knowing the term "social class," in general we would expect awareness of the term to be associated with awareness of the concept. In the Baltimore study, children were asked: "Have you ever heard about 'social class' or haven't you ever heard this term?" The results (Simmons and Rosenberg, 1971: 244) show that only 15 percent of the elementary school subjects had ever heard the term, compared with 39 percent of those in junior high school and 75 percent of those in senior high school. Furthermore, when those who said they *had* ever heard of social classes were asked which of four classes (upper, middle, working, or lower) they belonged to, 11 percent of the younger but 3 percent of the older said "don't know."

Equally important from the self-esteem viewpoint is the *accuracy* of the individual's perception of his own location in the stratification system. Assuming the extreme situation of a completely random association between objective class and subjective class identification, objectively high position would generate no feelings of pride nor objectively low position feelings of shame.

We shall draw upon data from three studies to examine whether children are in fact less likely to be aware of their actual status in society.

Table 5–7 presents the following data: (1) the association of objective class (Hollingshead scale) and subjective class identification among those Baltimore children and adolescents who had ever heard the term "social class" and who identified with one of the social classes; (2) the association between income and income class identification and between occupational status and prestige class identification among Chicago adults; and (3) the association between objective class (Hollingshead) and subjective class identification among Kohn and Schooler's (1969) national sample of adult working men.

**TABLE 5-7.**

*Objective and Subjective Class Identification, by Age*

| Association of objective and subjective class | Baltimore Children[a] | | | Chicago Adults | | Nation-wide male adults[d] |
|---|---|---|---|---|---|---|
| | Age | | | Prestige class[b] | Income class[c] | |
| | 8-11 | 12-14 | 15+ | | | |
| Gamma | −.0184 | .1972 | *.4412 | *.4912 | *.6935 | *.3954(r) |
| Number | (88) | (163) | (304) | (1234) | (1935) | (3074) |

[a] Among those who had ever heard of social class and who identified with a class.
[b] Occupational level and prestige class identification.
[c] Income and income class identification.
[d] Kohn and Schooler nationwide sample of 3,101 men using Hollingshead SES measure; figure is correlation coefficient.
*See footnote 7, chapter 3, p. 98.

The differences are striking. Among the youngest children, there is virtually no association (gamma = .0184) between their objective class and their social class identification. In the 12–14 age group, gamma = .1972; in the 15 or older group, gamma = .4412.[8] Three measures of association appear for adults. In the Chicago study, the occupation-prestige class identification association is gamma = .4912, and the income-income class identification association is gamma = .6935. Finally, the Kohn and Schooler study of 3,101 working men shows the correlation of objective class and subjective class identification is r = .3954. If the Pearson correlations of objective and subjective class among the school age groups are compared with each of the three adult correlations, they all prove to be significantly different at the .01 level. The correlations for the 8–11 and 12–14 age groups are also significantly different from the 15+ age group at the .02 level, but are not significantly different from one another.

Adults thus appear to be far more accurate than children about their relative status position in society. In particular, they are highly sensitive to income differences; the rich do know they are richer, the poor, poorer. Although there is variation in results, depending upon the samples and indicators used, in general among adults objective position does bear a strong and clear relationship to where one places oneself in the stratification system.

Like the other principles we have dealt with, the principle of psychological centrality applies just as strongly to children as to adults. The objective fact of social class, however, has a different phenomenal significance for adults and children and, hence, different self-esteem consequences. To the child, the social class system and his place in it is relatively vague, inaccurate, and peripheral; hence, objective class has little bearing on his self-esteem. For the adult, the reverse is the case. The significance of the objective demographic fact thus depends on the place of that fact in the individual's self-concept structure.

Certain of these principles also help us to understand Kohn's finding that among adult working men the relationship between social class and self-esteem is due largely to the degree of occupational self-direction. The principle of self-attribution would lead us to expect that someone who sees the visible outcome of his efforts, the products of his own decisions, would feel greater respect for himself than someone who does what he is told and cannot attribute the results to himself. Furthermore, the person who exercises occupational self-direction is likely to find work more psychologically central than the worker who, unable to exercise initiative or use his abilities, centers his feeling of self-respect in areas external to the world of work.

## Discussion

In directing attention to the fact that the identical demographic classification may represent different sets of social experiences and be accorded different phenomenal interpretations, it is important to stress that we are not speaking of different cognitive processes, modes of conceptualization, or motives among children and adults. On the contrary, the general principles governing self-esteem formation among children and adults are, we believe, identical. But one cannot understand the significance of a social structural variable for the individual without learning how this variable enters his experience and is processed within his own phenomenal field. As both the phenomenologists and the symbolic interactionists remind us, facts must be interpreted within their "meaning contexts." If we hope to appreciate the meaning of social class (or, for that matter, the meaning of any social identity element, such as race, gender, or religion) for the child, it is essential to see social class from his viewpoint, to adopt the child's-eye view of stratification, to understand how it enters his experience and is internally processed. To the sociologist, social class means differential prestige, respect, possessions, and power, with obvious self-esteem implications. But from the viewpoint of the child, the matter appears entirely different. The child, as he looks around him in the actual world of school and neighborhood in which he lives, finds that most of the children he meets are socioeconomically much like himself—neither richer nor poorer. His effective interpersonal environment, which provides the primary social experiences which enter his phenomenal field, is largely a classless society, a world in which status plays little or no role. (This is not to say that it is unstratified—age and authority stratification are sharp and strong—but that it is not *socioeconomically* stratified.) Again, the child sees himself through the eyes of others, and what he believes they think of him largely affects his self-esteem. And, indeed, significant others do think well or ill of him in many ways—but not with reference to socioeconomic status. Furthermore, even if the child knows where he stands, he is surely aware that this position has nothing to do with his own efforts or accomplishments. It is certainly understandable that he should feel greater pride or shame in his school marks or athletic skill than in paternal achievements. Finally, he is relatively oblivious to the social class system. Since his actual social class position bears only a modest connection with the class with which he identifies, one can understand why the young child is unlikely to take pride in high status or experience shame in low status. The differential association of social class to self-esteem for children and adults stems from the different social experiences and psychological interpretations associated with this structural fact in these age groups.

# NOTES

1. These data were collected by Leonard I. Pearlin. The sample was drawn in clusters of four households per block, and used a total of 575 blocks, one-fourth the total sample of 2,300. The 1970 census reports that there are 2,137,185 households in the Chicago urbanized area (U.S. Bureau of the Census, 1972); when this total is

divided by the total number of blocks in which households were to be chosen (575), the result, 3,716, is the skip factor for the selection of households. That is, every 3,716th household was selected; the three additional households of the block cluster were then chosen by dividing the total number of households on each block by four and using the result as the factor for counting from the initially selected address. Among those contacted, 30 percent refused to be interviewed. In anticipation of refusals and to make allowance for households where contact could not be established within three callbacks, substitute addresses in each block were prelisted. The targeting of households was thus entirely separated from the interviewing.

2. A suppressor variable is one which operates to conceal the true strength of the zero-order relationship. Whereas first-order extraneous and intervening variables reduce the size of the original zero-order relationship, first-order suppressor variables increase the size of this relationship. In the Baltimore study, the black children were more likely to be in the *lower* class but also more likely to have *higher* self-esteem. The fact that this test factor was negatively related to the independent variable but positively related to the dependent variable obscured the true relationship between social class and self-esteem in this sample. Hence, we have partialed out the effects of race.

3. We wish to thank Dr. Melvin Kohn for preparing this and several other special tabulations for inclusion in this chapter.

4. Kohn's (1969) and Kohn and Schooler's (1969) two factors of self-confidence and self-deprecation are based on 6 of the 10 RSE items. Kohn (1969: 84) showed that when occupational self-direction among adult working men was controlled, the proportional reduction in the correlation of social class to self-confidence was .99, and in the correlation of social class to self-deprecation, it was .87. One cannot, of course, be absolutely certain that this reduction in the social class self-esteem relationship is due solely to occupational self-direction and not at all to some related variable, but there can be little doubt that the job imperatives play a major role.

5. These calculations exclude those who considered their incomes equal to those of friends, relatives, or neighbors.

6. It may be noted that informal leaders are also more likely than others to have high self-esteem (Rosenberg, 1965: 26).

7. Such valuation of self-concept components, of course, is no mere accident. People not only attempt to be successful at those things they value, but they also elect to value those things at which they are successful. The point is that regardless of the factors which operate to make a disposition or social identity element psychologically central in the first place, once the element acquires that location in the individual's phenomenal field, its impact on the individual's global feeling of self-worth is amplified.

8. The developmental trend observed in Baltimore is reproduced in other research. Stern and Searing (1976: 185) note: "By the end of secondary school, family social class correlates with self-selected social class labels at gamma = .46 . . . in England and gamma = .58 in the United States"—results consistent with the gamma = .44 among Baltimore adolescents.

# 6

---

# Minority Status and Self-Esteem: An Inquiry into Assumptions

---

THAT PREJUDICE, not only against blacks, but also against many other minority groups—Jews, Mexicans, Japanese, Chinese, Catholics—has existed and continues to exist in American society is a dismal fact of history. The implications of this fact have, to most people, appeared self-evident. If a group is disdained in a society and its members treated with contempt, then the principle of reflected appraisals would lead one to expect members of that group to see themselves accordingly, that is, to have low self-esteem.[1]

Because blacks have borne the heaviest burden of prejudice and discrimination in American society, most research and theory has centered on this group. A perusal of the literature indicates that until recently the assumption that blacks had lower self-esteem appeared to be almost universal (Clark and Clark, 1947; Clark, 1965; Proshansky and Newton, 1968).[2] Similar views have been presented regarding the self-esteem of other discriminated-against groups, such as Mexican Americans, Puerto Ricans, Jews, and American Indians.

If one examines these various theoretical formulations more care-fully, however, most of them will be found to rest largely on two of the concepts which have been constantly invoked as central explanatory principles in the course of this work: reflected appraisals and social comparisons.

The principle of reflected appraisals, including both direct reflections and the generalized other, would lead us to expect lower self-esteem among members of socially derogated racial, religious, or ethnic groups. Since the self-concept is largely built up by adopting the attitudes of others toward the self, it follows that if others look down on us by virtue of our racial, religious, or ethnic group, in time we will come to see ourselves more or less as they do.

Unlike the concept of direct reflections, that of the generalized other, as pointed out in chapter 2, does not refer to other people's attitudes toward us. Rather, it maintains that, as a consequence of taking the attitudes of particular others toward the self, in the long run we come to internalize the viewpoints or perspectives of the society as a whole toward the self, particularly in structured role relationships. There is no reason to think that the individual's racial, religious, or national statuses are excluded. Just as we internalize the attitudes of the society as a whole toward us in terms of our sex, age, or occupation (as well as with reference to specific norms), so we should internalize their viewpoints toward us in terms of our race, religion, or nationality.

The social comparison principle should, on theoretical grounds, lead to the same conclusion. Minority group members, it is held, have lower self-esteem because they compare unfavorably with the majority group in ways other than their group membership. These unfavorable comparisons, such as low social class position, poor school performance, stigmatized family structures, are themselves consequences of prejudice and discrimination. But, minority group membership aside, the individual will use them as bases for comparing himself with others, and these comparisons can only be damaging to his self-esteem. Thus, the black may have low self-esteem not because he is black but because he is more likely to be an occupational failure; he is more likely to be an occupational failure, of course, because of prejudice and discrimination against blacks.

In sum, both persuasive theory (reflected appraisals and social comparison processes) and widespread popular and scholarly opinion converge in concluding that members of derogated groups will have lower self-esteem. In fact, everything stands solidly in support of this conclusion except the facts—at least those facts yielded by the relatively large-scale systematic surveys of the sixties.[3] Admittedly the data are extremely uneven and, of course, the life situations of each minority group are unique. Keeping in mind these limitations, it is relevant to consider what some of the research has to say regarding the relationship of minority status to self-esteem.

## Minority Group Status and Self-Esteem

*General Group Ranking.* The New York State study examined the self-esteem of various ethnic or racial groups within religious group categories.[4] Table 6–1 presents the distribution of self-esteem among 14 ethnic and racial groups (including mixed categories of "others"). Because ethnicity is related to social class, the data presented are controlled on social class by means of test factor standardization (Rosenberg, 1962b). Several points may be noted:

1) First, there is no indication that the distribution of self-acceptance in a group is related to the social prestige of that group in American society. In Table 6–1, we see that blacks, who are exposed to the most intense, humiliating, and crippling forms of discrimination in virtually every institutional area, do not have particularly low self-esteem.

They are, indeed, below average, but not by a conspicuous margin (45 percent of the total sample have high self-esteem compared to 39 percent of the black youngsters). At the same time, adolescents of English or Welsh descent, who certainly represent the essence of Old Yankee stock and whose pride is buttressed by a long tradition, are also slightly below average in self-esteem. For the other groups, the level of self-esteem within the group shows no relationship to the prestige rank accorded them in the broader society. For example, we have compared the Bogardus (1959: 441) attitudes toward ethnic groups with the proportion of corresponding ethnic groups (represented by 25 or more cases) with high self-esteem. The Spearman Rank Correlation coefficient is .04, indicating virtually no correlation.[5]

**TABLE 6-1.**

*Ethnic or Racial Groups and Self-Esteem,
by Religion (New York State)*

| | Self-Esteem | | | Total | |
| --- | --- | --- | --- | --- | --- |
| | High | Medium | Low | Percent | Number |
| **Catholics** | | | | | |
| German Catholics | 48% | 27 | 25 | 100 | (64) |
| Italian Catholics | 45% | 25 | 30 | 100 | (643) |
| Irish Catholics | 39% | 25 | 36 | 100 | (120) |
| Spanish-Portuguese Catholics | 28% | 32 | 40 | 100 | (25) |
| Polish Catholics | 28% | 28 | 45 | 100 | (65) |
| †All other Catholics | 44% | 26 | 30 | 100 | (998) |
| **Jews** | | | | | |
| German Jews | 62% | 19 | 19 | 100 | (21) |
| Russian Jews | 59% | 21 | 21 | 100 | (63) |
| Polish Jews | 51% | 20 | 29 | 100 | (35) |
| †All other Jews | 52% | 24 | 24 | 100 | (474) |
| **Protestants** | | | | | |
| German Protestants | 53% | 21 | 26 | 100 | (150) |
| English-Welsh Protestants | 39% | 24 | 37 | 100 | (122) |
| Black Protestants | 39% | 28 | 34 | 100 | (80) |
| †All other Protestants | 43% | 25 | 32 | 100 | (1375) |

†Includes those of mixed national origin, i.e., either (1) father and
mother are of different national origins, or (2) father or mother is of
mixed national origin.

*Religion.* That anti-Semitism is a firmly grounded feature of American life is a much remarked and well-documented fact. In terms of self-esteem data, the evidence is rather slim, but what is available affords little support for the assumption of lower self-esteem among Jews. Of the small number of studies on this subject, many show the Jewish subjects to have somewhat *higher* self-esteem (for example, Gordon, 1963; Anisfeld, Bogo, and Lambert, 1962; Bachman, 1970).

The New York State and Baltimore studies agree with these findings in general, though not in every particular. Table 6–2 (New York State) shows that not only do Jews have higher self-esteem than that of Catholics or Protestants, but that there is no difference in the self-esteem of Catholics or Protestants, despite the traditional differences in the prestige rank of the two religious groups.

**TABLE 6-2.**

*Religion and Self-Esteem*
*(New York State)*

| | Religion | | |
|---|---|---|---|
| | Catholics | Protestants | Jews |
| *Self-Esteem* | | | |
| High | 43% | 43% | 53% |
| Medium | 26 | 25 | 23 |
| Low | 31 | 32 | 23 |
| N = 100% | (1727) | (1913) | (592) |

*Jews vs. Catholics and Protestants: proportion "High."

*See footnote 7, chapter 3, p. 98.

In the Baltimore sample, only a handful of the Jewish children attended elementary school. Hence, Table 6–3 deals only with junior and senior high school white children. Since all children at this level completed both the Rosenberg Self-Esteem (RSE) measure as well as the Baltimore (RSSE) measure,[6] it is possible to compare the self-esteem of the three religious groups in terms of both measures. The data show that, according to the RSE, there is little difference in self-esteem among the three major religious groups (Catholics are slightly less likely to be high in self-esteem, slightly more likely to be medium). On the Baltimore measure, Catholic self-esteem is distinctly higher.[7] Whatever one may wish to conclude from these data, neither study supports the assumption of a direct relationship between religious status and self-esteem.

*Race.* The assumption of low self-esteem among blacks is particularly compelling because of the multiplicity of disadvantageous life conditions which are the sorry heritage of American racism. The facts, however, are otherwise, at least among school populations. In his junior college study, Gordon (1963) found blacks to have the highest self-esteem of five groups. McDonald and Gynther's (1965) investigation of 261 black and 211 white high school seniors showed blacks to have higher self-esteem (based on self-ideal discrepancy and on dominance scores). Coleman's (1966) nationwide study showed no black-white difference in "academic self-concept," but McDill's (McDill *et al.*, 1966) sample of 327 high school blacks

**TABLE 6-3.**

*Religion and Self-Esteem, Among*
*Baltimore White Children*
*Seventh Grade or Higher*

| | Rosenberg Self-Esteem Scale (RSE) | | |
| --- | --- | --- | --- |
| | Protestant | Catholic | Jewish |
| *Self-Esteem* | | | |
| High | 45% | 40% | 45% |
| Medium | 22 | 31 | 25 |
| Low | 33 | 29 | 30 |
| N = 100% | (162) | (108) | (56) |
| | Baltimore Self-Esteem Scale (RSSE) | | |
| High | 28% | 45% | 30% |
| Medium | 33 | 23 | 26 |
| Low | 39 | 32 | 43 |
| N = 100% | (158) | (102) | (53) |

individually matched with whites showed higher academic self-concepts among blacks. Large sample studies by Hunt and Hardt (1969), Powell and Fuller (1973), and Bachman (1970) all showed blacks with higher self-esteem, and this difference usually increased when class, IQ, and family structure were controlled.

The data for adults are more sparse, but the conclusions which can be drawn are not discordant. Middleton (1972) showed adult black self-esteem to be lower than white but, when socioeconomic variables were controlled, to be significantly higher. Kohn's (1969) nationwide study of 3,101 adult males showed blacks to be non-significantly more self-deprecatory but significantly more self-confident than whites. Yancey, Rigsby, and McCarthy (1972), in their study of 1,179 adult subjects in Nashville and Philadelphia, found that when marital status, work force participation, sex, age, and education were controlled, race was related to self-esteem in opposite ways in the two cities.

The most thorough review of this literature has been conducted by Ruth Wylie (1978). In this work, Wylie has reviewed 53 publications dealing with the relationship between racial or ethnic status and global or specific self-esteem. Viewing all these studies together, she concluded that "the results . . . place the burden of proof on those who have contended that the derogated, disadvantaged social position of the blacks in the United States must obviously have resulted in seriously damaged self-esteem in that group."

*Ethnicity.* Social scientists and educators have recently turned their attention to the plight of the Mexican-Americans. Unlike the black self-esteem data, which are abundant, the evidence with regard to Mexican-Americans is scattered and fragmentary. Healey and DeBlassie (1974) studied 630 ninth graders in two junior high schools in a south-central New Mexico town. The sample, consisting of 425 Anglos, 40 blacks, and 142 Spanish-Americans, was similar to the population distribution for the state as a whole. The self-esteem measure employed was the Total Net Positive Score of the Tennessee Self-Concept Scale. Contrary to expectations, the results showed the Spanish-American youngsters to be highest in self-esteem, blacks to be second, and Anglos third. The differences, however, were not significant.

Carter (1968) also studied ninth graders (190 Mexican-Americans and 98 Anglos). Subjects were requested to rate themselves in terms of intelligence, goodness, happiness, and power, using the semantic differential. There was little difference between the two groups in terms of these traits. Of special interest is the fact that when Carter interviewed teachers and administrators in these schools, they were convinced that the Mexican-American children did have lower self-esteem and offered a number of persuasive reasons to support their erroneous conclusions.

Guzman (1976) has reviewed 9 studies comparing the self-esteem of Mexican-Americans with Anglos. Hishiki (1969), Gillman (1969), McDaniel (1967), Peterson and Ramirez (1971), and Coleman *et al.* (1966) showed lower self-esteem among Mexican-Americans, whereas Carter (1968, 1970), Healey and DeBlassie (1974), Najmi (1962), and Cooper (1971) showed Mexican-American self-esteem to be as high as, or higher than, Anglo self-esteem. It is difficult to draw any general conclusions from these studies because of methodological limitations of various sorts.

Even less is known about Puerto Rican self-esteem. Zirkel and Moses (1971) compared the self-esteem of 120 fifth- and sixth-grade Puerto Ricans, blacks, and whites in three Connecticut schools. The groups were balanced in terms of sex, SES, and IQ. Using the Coopersmith SEI, the Puerto Ricans rated significantly lower than the blacks or whites.

On the other hand, Guzman (1976) studied 307 black, white and

Puerto Rican children in four schools (selected for group context) in Buffalo, New York. The groups were similar in terms of SES, sex, and IQ. Using the Piers-Harris Children's Self-Concept Scale, blacks had the highest self-esteem, Puerto Ricans second, and whites lowest.

Firm conclusions about whether Spanish-Americans do or do not have lower self-esteem than Anglos cannot be drawn at this point. Certainly there is no unequivocal evidence to indicate that the considerable prejudice and discrimination which has been their lot has generated a corresponding self-esteem level in its members.

When we turn to the self-concepts of American Indians (native Americans), systematic research is hard to find. A study by Trimble (1974), however, strongly casts doubt upon the assumption of low self-esteem among American Indians. Trimble conducted a survey on five reservations in selected Indian communities in Oklahoma and at Haskell Indian Junior College in Lawrence, Kansas. A total of 1,110 persons (a random selection of about 120 interviews in each area) were surveyed by bilingual interviewers. The data showed high self-esteem levels among these respondents. Obviously, a single study, even one of relatively high quality, does not suffice to establish an empirical generalization, but it certainly lends no support to the assumption of low self-esteem among members of this disprivileged group.

## SUMMARY

As one reviews the available evidence regarding the relationship of "ethnic" or "minority group" status (such as race, religion, or ethnicity) to self-esteem in school populations, it is difficult to see how it supports the oft-asserted view that members of these ethnic groups have lower self-esteem.[8] This evidence, to be sure, is beset with numerous limitations: for many ethnic groups, there are almost no data; for other groups, only a few studies are available; and even in the most richly investigated area of race, there are limitations and variations of sample and method. Furthermore, it is possible that changes have occurred over time. Yet no reasonably dispassionate review of the currently available evidence can justify the conclusion that minority group children have lower self-esteem. The overwhelming body of evidence suggests that they do not.

Since the theories of reflected appraisals and social comparison lead to one conclusion while the data lead to another, the question is: which is wrong—the theories or the data? The answer is: neither. The explanation, we believe, is to be found in the fact that many scholars, as they apply the theories to the given case, make certain implicit assumptions which, when exposed to explicit examination, turn out to be largely unjustified. Our first task will be to examine certain assumptions underlying the application of the reflected appraisals principle, reserving consideration of the social comparison principle for later discussion. It should be kept in mind that the analyses and reasoning presented here apply to children (between 8–18). In some respects, it is certain, the experience of minority group membership will differ for children and adults and might well have a different impact on self-esteem.

## Reflected Appraisals

Whether reflected appraisals are conceived of as direct reflections or in terms of the generalized other, both conceptions are consistent in predicting that people who are derogated in the society as a whole because of their membership groups will, in the long run, come to internalize society's negative attitude toward them. Yet consider blacks as a prototypical group. Since the evidence showing negative feelings toward blacks in our society is clear, and is equally clear in showing satisfactory self-esteem among blacks (at least among children), it might appear that the theory is at fault. We do not believe this is so. If one considers the matter more carefully, it becomes evident that the conversion of society's attitude toward one's group (including oneself as a group member) into the individual's attitude toward the self is logically compelling only if certain assumptions are sound. The first is that the individual knows how the broader society feels about his group (the assumption of *awareness*). The second is that he accepts the societal view of his group (the assumption of *agreement*). The third is that he believes the societal view of the group's characteristics apply to the self (the assumption of *personal relevance*). And the fourth is that he is critically concerned with the majority attitudes (the assumption of *significance*).

We believe there is both logical and empirical reason to challenge each of these assumptions, in part or in whole.

## THE ASSUMPTION OF AWARENESS

The assumption of awareness would appear to be the least questionable of all. How can anyone live in this society, with its intricate, elaborate, and conspicuous system of prejudice, and not be clearly aware of what the society thinks of his group? Part of the answer, stressed in our discussion of social contexts, lies in the fact that no one ever lives in the total society but only in a limited segment of it, and that a variety of powerful social forces operate to insure that this special segment is not representative of the whole.

This is particularly true of school-age populations. In the Baltimore study the children were instructed to indicate which of "four different kinds of Americans . . . MOST PEOPLE in America think is best: Jewish; white Catholic; Negro; white Protestant." Observe that children are asked which group "most people in America" think is best, not which *they* think is best. Nevertheless, the black children show a striking tendency to believe their race is judged favorably. Table 6–4 shows that, for the sample as a whole, fully 43 percent of the blacks, but just one-half of one percent of the whites, believe

**TABLE 6-4.**

*Children's Opinions of "Most People's" Ranking of Blacks, by Race and Age*

| | 8-11 years | | 12-14 years | | 15+ years | | Total | |
|---|---|---|---|---|---|---|---|---|
| | Blacks | Whites | Blacks | Whites | Blacks | Whites | Blacks | Whites |
| Of four racial-religious groups, "most people" rank blacks: | | | | | | | | |
| First | 63% * | 1% | 32% * | 1% | 15% * | – % | 43% * | 1% |
| Second | 17 | 6 | 26 | 4 | 14 | 1 | 18 | 4 |
| Third | 9 | 24 | 20 | 24 | 26 | 17 | 17 | 23 |
| Last | 11 | 69 | 23 | 71 | 44 | 81 | 22 | 72 |
| N = 100% | (425) | (172) | (243) | (212) | (250) | (149) | (996) | (557) |
| Difference in percent ranking blacks "Last" | 58 | | 48 | | 37 | | 50 | |

*See footnote 7, chapter 3, p. 98.

blacks rank highest in status. Indeed, fewer than 1 out of 20 whites, compared with 12 out of 20 blacks, believe blacks are located in one of the two top ranks. Conversely, 72 percent of the former but 22 percent of the latter, believe blacks are ranked by "most Americans" as lowest in prestige.

Such selective attribution, it should be noted, is not confined to the black children in the sample, but obtains strongly among the white children as well. Table 6–4 indicates that the Protestants are more likely than the non-Protestants to believe their group ranks high; the Catholics are more likely than non-Catholics to think their group ranks high; and the Jews are more likely than gentiles to believe their group ranks high.

No one is really likely to challenge these findings, but many may challenge the interpretation that they reflect actual lack of awareness rather than an unwillingness to accept an unpalatable truth which is actually recognized at an unconscious level. This objection deserves special comment.

First, there is ample evidence to indicate that people respond to reality only insofar as it intrudes on their experience. Certainly, as black children grow older and their knowledge and experience broaden, they become increasingly accurate in their perceptions of society's attitudes toward their group. This is clearly shown in Tables 6–4 and 6–5; the older the child, the more accurate the perception of his group status.[9]

Second, as we saw in the discussion of contextual dissonance, both black and Jewish children attending schools in which they were in the minority are more likely to report that they have experienced racial or religious slurs or exclusion on grounds of their group membership. It is thus significant to note that those black children attending desegregated schools are also more likely than those in segregated schools to believe their group is ranked low in the society. If the older and the desegregated children are more aware of societal prejudice, it is not because their motive to protect their self-esteem is any weaker, but because their experience has forced them to be aware of the truth, unpalatable though it may be.

The idea that minority children are unwilling to confront the truth is thus an oversimplification. No one ever responds to the truth; he only responds to the truth as it enters his experience. Since, as we observed in the contextual dissonance chapter, children tend to interact primarily with members of their own group, it is under-

## TABLE 6-5.

White Children's View of "Most People's" Ranking of Religious Groups, by Age

### "Most people" rank . . . Protestants:

| | 8-11 years | | 12-14 years | | 15+ years | | Total | |
|---|---|---|---|---|---|---|---|---|
| | Protestant | Non-Protestant | Protestant | Non-Protestant | Protestant | Non-Protestant | Protestant | Non-Protestant |
| First | 60% * | 15% | 54% * | 22% | 47% | 52% | 55% * | 32% |
| Second | 33 | 69 | 38 | 66 | 45 | 38 | 38 | 56 |
| Third | 6 | 9 | 6 | 9 | 6 | 10 | 6 | 10 |
| Last | 1 | 7 | 2 | 2 | 2 | – | 1 | 2 |
| N = 100% | (125) | (54) | (125) | (98) | (64) | (94) | (314) | (247) |

### "Most people" rank . . . Catholics:

| | Catholic | Non-Catholic | Catholic | Non-Catholic | Catholic | Non-Catholic | Catholic | Non-Catholic |
|---|---|---|---|---|---|---|---|---|
| First | 88% * | 45% | 93% * | 46% | 57% | 44% | 80% * | 45% |
| Second | 9 | 46 | 7 | 45 | 43 | 56 | 19 | 48 |
| Third | 3 | 7 | – | 7 | – | 1 | 1 | 5 |
| Last | – | 3 | – | 2 | – | – | – | 2 |
| N = 100% | (34) | (148) | (55) | (166) | (42) | (117) | (132) | (431) |

### "Most people" rank . . . Jews:

| | Jew | Non-Jew | Jew | Non-Jew | Jew | Non-Jew | Jew | Non-Jew |
|---|---|---|---|---|---|---|---|---|
| First-Second | – | 12% | 27% | 12% | 22% * | 5% | 26% * | 10% |
| Third | – | 63 | 73 | 62 | 74 | 73 | 72 | 65 |
| Last | – | 26 | – | 27 | 3 | 22 | 2 | 25 |
| N = 100% | (1) | (171) | (11) | (202) | (31) | (118) | (43) | (492) |

standable that the child should not be completely aware of the attitudes of the society as a whole toward his group. And, if he is relatively unaware, then neither direct reflections nor the generalized other should substantially damage his self-esteem.

### THE ASSUMPTION OF AGREEMENT

But even if we *know* what the rest of society thinks of our group, does that mean we *agree* with their judgment? Although the principle of reflected appraisals suggests that we tend to see ourselves as others see us, this generalization must be qualified as follows: *we are more likely to agree with their views if those views are favorable than if they are unfavorable.*

Consider again the study by Reeder, Donohue, and Biblarz (1960) asking soldiers to rate both themselves and other soldiers in their group in terms of leadership. Of 11 soldiers rated by their peers as high in leadership, 10 rated themselves as high; but of 31 soldiers rated by others as low, only 14 rated themselves low. The soldier is more likely to agree with the group judgment of what he is like if that judgment is favorable than if it is unfavorable.

Since society's views of a low status group are by definition negative, it is unlikely that group members will generally agree with the societal judgment. But there are several different types of negative assessments. One is to assign the derogated group low prestige or status relative to other groups in the society; the second is to attribute to the group negative stereotypes or to consider the characteristics attributed to the group to be negative; and the third is to consider the group less adequate or worthy than another in some valued respect. We shall consider these three types of social derogation separately.

*Group Prestige.* As far as judgments of group prestige are concerned, two conclusions appear warranted. The first is that in general all groups—discriminators and discriminated against, high or low—share a common system of racial, religious, and occupational prejudice. Minority and majority group members alike learn and internalize the same general values and attitudes toward minorities. The second point is that, although groups generally agree on the rank order of prestige in the society, they pull their own group out of the order and accord it a higher status than others do. A synthesis of the relevant literature has been provided by Harding *et al.* (1969: 23–24), who note that there is a "relatively enduring and uniform hierarchy of preferences for ethnic groups in America." They note:

In the studies by Zeligs and Hendrickson (1933, 1934) in Cincinnati, it was found that when 39 ethnic groups were ranked in order of preference, the correlations between the ranking by Jewish and non-Jewish children was quite high. Other investigators have consistently found high correlations between the rankings of minority and majority group high school and college youth (Gray and Thompson, 1953; Horowitz, 1944; Katz and Braly, 1935; Meltzer, 1939a, 1939b). The only notable exception to the pattern of agreement in rankings among different groups occurs when group members rank their own group. In this instance a difference arises because each group ranks itself as the most preferable.

*Stereotypes.* What holds true for general group status is equally true for group stereotypes. Negative stereotypes of the group can give rise to low self-esteem *only if the individual accepts these stereotypes as true* and as negative. A study of Brigham (1974) asked black and white segregated school subjects to rate a series of 50 "stereotype-relevant" traits as more characteristic of blacks, more characteristic of whites, no difference, or "don't know." Ten months later some of the subjects were asked to indicate how likeable each trait was. According to Brigham (1974: 154): "For this sample of traits, both groups showed a significant tendency to attribute favorable traits to their own race, but this tendency was significantly greater for the black children than it was for the white children." Thus, the self-esteem of both groups was enhanced.

This proposition can probably be accepted as universal. A study of intergroup attitudes among groups in East Africa (Brewer and Campbell, 1976) asked subjects to rate their own groups and others in terms of a diversity of traits. In virtually every case, group members rated their own groups most favorably.

*Superiority-Inferiority.* Whereas group prestige refers to the general rank in the society of a number of racial, religious, or ethnic groups, group superiority-inferiority refers to which of two groups are more or less meritorious in some particular respect. Two studies may be mentioned in this regard. The first is that of Laurence (1970), who examined 178 black and 821 white children in fifth, sixth and eighth grade integrated classes in Sacramento. The children were asked to *compare* blacks and whites in terms of three characteristics: behavior; intelligence; and honesty. Table 6–6 shows that virtually no whites considered blacks "better" on any of these qualities; but, not many black children considered whites better, either. The pattern that appears to emerge is this: some fairly substantial proportion of the white children—usually about a third—say that whites are better,

**TABLE 6-6.**

*Comparative Evaluations of Negro
and White Characteristics*

| Characteristics | Percentage Distributions[†] | | | |
| | Young | | Old | |
| | White | Black | White | Black |
| --- | --- | --- | --- | --- |
| Behavior | | | | |
|   Whites behave better | 35 | 9 | 43 | 10 |
|   Negroes behave better | 1 | 13 | – | 7 |
|   Both are about the same | 47 | 63 | 45 | 66 |
| Intelligence | | | | |
|   Whites are smarter people | 22 | 6 | 30 | 9 |
|   Negroes are smarter people | 1 | 12 | 1 | 15 |
|   Both are about the same | 57 | 65 | 53 | 57 |
| Honesty | | | | |
|   Whites are more honest | 33 | 5 | 35 | 7 |
|   Negroes are more honest | 1 | 15 | 2 | 13 |
|   Both are about the same | 41 | 58 | 43 | 50 |
| Number of respondents | (408) | (110) | (413) | (68) |

†Totals for each characteristic do not equal 100 percent because
"don't know" responses are not presented.

    Source: Joan Laurence. "White socialization: black reality."
*Psychiatry: Journal for the Study of Interpersonal Processes* 33,
1970: 174-194, Table 6. (Reprinted with permission. Copyright ©
1970 by The William Alanson White Psychiatric Foundation, Inc.)

but only a small proportion of the blacks agree with them. In other
words, many whites consider whites superior, but most blacks consider blacks equal.[10] Essentially the same results appear among both
younger (mostly 10–11 years) and older (mostly 12–13 years) subjects.

    Russell Middleton's (1972) results, though not identical in every
particular, are essentially similar. He asked: "Do you think that
Negroes or whites tend to have more inborn intelligence or are they
both about the same?" "Who do you think tend to be lazier and
less ambitious (more dishonest; have looser sexual morals; keep
their homes clean, neat, and attractive)—low income Negroes or
low income whites?" Although there are certain puzzling regional
differences, the data bear out the obvious conclusion that blacks and
whites differ radically in their perceptions and judgments. Blacks do
not agree at all with those whites who consider them inferior in one
or another regard.

In sum, the data offer little support for the assumption that minority group members agree with society's negative attitudes toward their group, whether expressed in low group prestige, negative stereotypes, or claims of inferiority. It is true that we live in a prejudiced society, and that a substantial proportion of the white population holds negative stereotypes of blacks or considers them inferior in some way. What is probably not true—and this is crucial from the self-esteem standpoint—is that any appreciable number of blacks attribute traits to their race which they consider unfavorable or believe that blacks are generally inferior to whites. In fact, according to Wylie (1978), "recent evidence suggests a greater group-favorability bias in blacks' attributions to blacks in general than to whites' attributions to whites in general."

The reasons are obvious. The individual's intimate contact with members of his own group makes the complete acceptance of these stereotypes a virtual impossibility. It is possible for a white person to believe that all blacks are lazy, ignorant, and superstitious (Katz and Braly, 1935), but no black can possibly believe this to be the case, for it obviously contradicts his direct and immediate experience.

Another reason why minority group members—children in particular—would be unlikely to share the general societal judgment or evaluation with regard to their groups is the "communications environment," discussed in chapter 4. Although the child may absorb from the adults in his effective interpersonal environment (for example, his parents) the general attitudes toward such issues as honesty, respect for property, sex role assignments, or even the comparative "goodness" or prestige of races, religions, or other evaluated statuses, it is extremely unlikely that he will absorb from these particular others the view that the negative attitude of the generalized other toward his own group is a valid one. Instead, from these particular others, the child is more likely to hear about the admirable qualities of his group; conversely, the negative stereotypes associated with the group are more likely to be denied, defended, or explained. Hence, each child is likely to emerge with a particularly favorable view of his own group. Although the negative stereotypes or lesser worth of *other* groups may be accepted as gospel, those applied to one's own group will be rejected as libel.

Finally, it is important to keep in mind that the minority individual is not a blank stencil accepting every imprint with which society strikes it. He deals in an active and creative way with the social stimuli to which he is exposed, guided by a powerful system of motives in which the self-esteem motive holds a prominent place. Thus, even if the black, for example, were fully aware of the low repute in which his group is held, this awareness does not compel him to concur. What the black may correctly conclude is that he lives in a bigoted, irrational society. But it is the white who is bigoted and irrational, not the black race in general, nor himself in particular, who is at fault. Hence, to know what whites think about his race does not oblige him to share this view.

THE ASSUMPTION OF PERSONAL RELEVANCE

Assume, however, that the individual *is* aware of what the broader society thinks of his group and *does* agree with society's judgment of it. Will he therefore conclude that, since he is a group member, these negative stereotypes necessarily characterize *himself*? Although this reasoning is common,[11] the obvious flaw lies in the assumption of personal relevance—that since a person is a member of a minority group, he will assume that the group characteristic applies to the self. But plainly this is not the case. The individual black who considers most blacks crude or ignorant may take special pride in being precisely the reverse. The Jew who despises most Jews as loud, pushy, and bad-mannered (the classical case of Lewin's "Jewish self-hatred") may be especially proud of his own tact, manners, and cultivation. Obviously, it is possible for members of minority groups to agree with the stereotypes and broad judgments about their groups without believing that these characterize them personally. Under these circumstances, it is as likely that the individual will take special pride in his difference from his maligned group as experience personal shame at his similarity to it.

Age stereotypes afford a good case in point. A nationwide survey (Harris, 1975) compared the stereotypes of the aged among different age groups. The subjects were asked about seven characteristics of "most people over 65": friendly and warm, wise from experience, bright and alert, open-minded and adaptable, good at getting things done, physically active, and sexually active. In general, the negative stereotypes of the aged were strongest among the youngest group (18–24), continued to become more favorable with increasing age, and declined somewhat after age 65. Although older people are not as likely as younger ones to hold negative stereotypes of the "over 65" group, nevertheless they do accept many of the negative stereotypes of their age category.

But does such agreement oblige the older person to believe that these stereotypes characterize him personally? Obviously not. Among people over 65, only 29 percent said that "most people over 65" were "very bright and alert," but fully 68 percent said that *they* were very bright and alert. For "very open-minded and adaptable," the corresponding figures are 21 percent and 63 percent, and for "very good at getting things done," 35 percent and 55 percent. We can think of no reason to assume that this reasoning is any less powerful among blacks, Jews, Mexican-Americans, Puerto Ricans, or American Indians. Group members are certainly unlikely to accept the negative stereotypes of their groups; but, even if they do, they will generally not apply these stereotypes to themselves.

Unfortunately, the Baltimore study lacks information on the traits group members attribute to their own or other groups, so that it is not possible to assess how the child's view of his group affects his view of himself. It does, however, contain data on how group members characterize themselves as individuals. Do black and white children characterize *themselves* in accordance with the racial stereotypes traditionally associated with their groups in the society at large? In order to examine this question, the Baltimore subjects were handed a card with four broad bars differing conspicuously in height and labeled, respectively, "very much," "pretty much," "not very much," and "not at all." They were asked to indicate, by pointing to the appropriate bar, which best described themselves in terms of a number of traits. For example: "First of all, how SMART are you? Are you very smart, pretty smart, not very smart, or not at all smart?" (Child points to appropriate bar or uses corresponding word.) The same format was used to get at such characteristics as good-looking, truthful and honest, good at sports, well-behaved, work hard in school, get other kids to do what you want, boys like you, girls like you, and others.

Because age is sometimes related in a rather complex fashion to these questions, the data presented are controlled for age by means of test factor standardization (Rosenberg, 1962b). For simplicity of presentation, Table 6–7 simply sets forth the mean scores of blacks and whites. A low mean score indicates that the children feel this is "very much" or "pretty much" like them.

Generally, there is little difference in the ratings black and white children assign themselves with regard to these characteristics. Even a superficial glance at Table 6–7 reveals a point of deep psychological significance: that, with regard to a fairly wide range of traits, black children are every bit as likely as white children of a comparable age to rate themselves favorably. Nor is there any indication in the data that, with regard to the negative racial stereotypes (usually expressed in the language of traits) characterizing the American prejudice system, black children are any more likely to have incorporated them into their self-concepts than white children.

**TABLE 6-7.**

*Mean Scores of Black and White Children's
Evaluations of 21 Traits or Characteristics
(Standardized on Age)
(Low mean score indicates that the trait is
"very much" like them)*

| Trait or Skill | Blacks (N = 1108) | Whites (N = 674) |
|---|---|---|
| Smart | 2.05 | 2.03 |
| Good-looking | 2.28 | 2.29 |
| Get angry | 2.74 | 2.62 |
| Truthful or honest | 1.73 | 1.74 |
| Good at sports | 1.99 | 2.04 |
| Well-behaved | 1.79 | 1.85 |
| Work hard in school | 1.65 | 1.75 |
| Get other kids to do what you want | 2.71 | 2.39 |
| Boys like you | 1.96 | 2.01 |
| Girls like you | 1.82 | 1.99 |
| Helpful | 1.66 | 1.74 |
| Good at making jokes | 2.59 | 2.52 |
| Easy to get along with | 1.55 | 1.66 |
| Friendly | 1.49 | 1.61 |
| Shy | 2.68 | 2.68 |
| Good at working with hands | 1.83 | 1.93 |
| Religious | 1.94 | 2.08 |
| Able to stand up for rights | 1.78 | 1.83 |
| Tough, not afraid of a fight | 2.01 | 2.24 |
| Able to get along without help | 1.95 | 1.94 |
| Able to decide things for self | 1.98 | 2.01 |

It is thus evident that even if the minority group member agreed that his group was inferior or was characterized by negative stereotypes, this would not oblige him to conclude that *he* was inferior or possessed these negative characteristics. The same principle applies to the assumption of awareness. Even if minority children were fully aware of the majority group's attitude toward his group, this awareness would not require him to believe that these descriptions aptly characterized himself. Consider Table 6–8, which compares the self-esteem of black children who believe that "most Americans" rank their group "best" with those who think it is ranked second, third, or last. The results are striking: there is virtually no connection between judged status rank and self-esteem. How the black child sees society's evaluation of his group appears to be totally independent of his personal feelings about himself. And this is true at each age level.

TABLE 6-8.

*Self-Esteem by Black Children's View of*
*"Most People's" Ranking of Blacks*

| | View that "most Americans" rank Blacks | | | |
| | First | Second | Third | Last |
| --- | --- | --- | --- | --- |
| *Self-Esteem* | | | | |
| Low | 21% | 15% | 20% | 22% |
| Medium | 33 | 37. | 31 | 33 |
| High | 46 | 47 | 49 | 45 |
| N = 100% | (412) | (177) | (157) | (209) |

THE ASSUMPTION OF SIGNIFICANCE

The conclusion that the attitudes of particular majority group others or of the society as a whole (the internalized generalized other) will produce corresponding self-attitudes is also predicated on the assumption of significance. As we indicated in our discussion of interpersonal significance, the attitude of the other toward the self will have a substantially greater impact if the other's opinion matters greatly than if it does not. The principle of reflected appraisals must thus be qualified to include the questions: which others? how significant? and compared to whom?

We do not, in fact, know how important the broader society's attitude toward the minority child's group, or toward himself as a group member, actually is (valuation); nor do we know how much faith or trust the child has in the majority group judgment (source credibility). But it is clear that the majority view is not of exclusive importance to the child, nor is his faith in the majority view absolute. Indeed, it is virtually certain that the minority child's self-esteem is far more heavily influenced by what his mother, father, teachers, siblings, and classmates think of him than by what the broader society thinks of him. The relationship between the perceived attitudes of significant others and self-attitudes is powerful and unequivocal, and applies to minority and majority children in equal degree. This is clearly revealed in Table 6–9, which examines the relationship between a "significant others" score—based on the child's assessment of what his mother, father, teachers, kids in his class, best friend, and brothers and sisters think of him—and his own global self-esteem. Among the black children, 70 percent of those perceiving the attitudes of these significant others as favorable, but only 23 percent of those perceiving these attitudes as unfavorable, have high self-esteem; for the white children, the corresponding figures are 43 percent and 12 percent. These are among the strongest empirical findings in the Baltimore study. Minority children thus assuredly do tend to see themselves as others see them but only *if the opinions of these others are highly valued.*

Nor can we automatically assume that source credibility for the majority opinion is very high. Any minority group member knows that the deprecatory attitudes and negative stereotypes held about his group are based on ignorance, bigotry, and irrationality. It may be upsetting to know what the majority group thinks of our group, but this is not to say that these attitudes are *believed.* And, as we noted in the discussion of significant others in chapter 3, the impact

**TABLE 6-9.**

*Self-Esteem and Attitudes Attributed to Significant Others, by Race*

| | *Blacks | | | | *Whites | | | |
|---|---|---|---|---|---|---|---|---|
| | Attitudes of significant others perceived as: | | | | | | | |
| | Favorable | | | Un-favorable | Favorable | | | Un-favorable |
| | 1 | 2 | 3 | 4 | 1 | 2 | 3 | 4 |
| *Self-Esteem* | | | | | | | | |
| Low | 4% | 13% | 26% | 38% | 38% | 32% | 41% | 61% |
| Medium | 26 | 36 | 38 | 39 | 19 | 28 | 36 | 28 |
| High | 70 | 51 | 36 | 23 | 43 | 40 | 23 | 12 |
| N = 100% | (50) | (409) | (252) | (69) | (21) | (286) | (116) | (33) |

*See footnote 7, chapter 3, p. 98.

of others' attitudes toward the self will be much weaker if source credibility is low than if it is high.

That the valuation and source credibility of primary significant others is considerably stronger than that of the society as a whole thus seems beyond reasonable dispute. In this regard, we are aware of no data to suggest that the minority child's significant others—mother, father, friends—have any less love or respect for him than the majority child's.[12] In the Baltimore study, the black children are at least as likely as the white children to believe their significant others hold positive attitudes toward them. And the views of these significant others are just as important for the minority as for the majority child's self-esteem. One of the errors in thinking that societal prejudice will directly affect global self-esteem lies in failing to give due consideration to the interpersonal centrality of different others in the lives of individuals.

## SUMMARY

The fundamental principle of reflected appraisals (whether direct reflections or the adoption of the perspective of the generalized other) holds that we come to see ourselves as others see us. In general, empirical data strongly and consistently support this assumption. It is also true that the society at large holds minority groups and their members in low regard and tends to think of them in terms of negative stereotypes. And yet, as far as present evidence enables us to judge for school populations, minority group members apparently neither have low regard for themselves as a whole nor believe that the negative stereotypes characterize themselves personally. The reason that the premises do not lead to the conclusion, as is so widely assumed in the literature, is that they conceal certain hidden assumptions: the assumption of awareness; the assumption of agreement; the assumption of personal relevance; and the assumption of significance. We do not suggest, of course, that these assumptions are *invariably* untrue, that they apply to *no* minority group members. Indeed, for some they are reality, and the psychological consequences may be serious, indeed. What we do suggest, rather, is that they are not invariably true and that most of the assumptions are not even generally true. The fact that minority group members—especially preadults—have satisfactory self-esteem levels despite the prejudice and discrimination of the society thus does no violence to the principle of reflected appraisals, which remains sound; it does, however, show that certain conditions must be met if the principle is to hold.

## Social Comparisons

Another reason for expecting lower self-esteem among minority group members is that certain derivative consequences of prejudice and discrimination produce highly unfavorable comparisons with the privileged majority. Again, blacks will serve as the prototypical group. Consider, for example, the issues of income, academic performance, family structure, and skin color. In all four respects, according to the general value system of the society, the black child is decidedly inferior to the white. He is much more likely to be poor (black income still remains at about three-fifths of the white average); his academic performance is far below the white (Coleman *et al.*, 1966); he is substantially more likely to stem from a stigmatized family structure—never-married or separated; and he is characterized by physical features which, *in the view of both blacks and whites,* are less aesthetic than the caucasoid model. (Evidence for this last statement will be presented shortly.) All these factors, themselves secondary consequences of racism, can independently affect self-esteem. Race aside, we would expect poor people to have lower self-esteem than rich ones; the academically unsuccessful student to have lower self-esteem than the academically successful one; the illegitimate or father-abandoned child to have lower self-esteem than one from an intact family; and for the darker child to have lower self-esteem than the lighter child. And, it should be pointed out, there is good evidence in support of every one of these predictions.[13] *Direct* prejudice aside, then, the social comparison principle would clearly suggest lower self-esteem among black children.

Since the principle of social comparison is sound, and since black children compare unfavorably with whites in a number of specific respects which are critical for self-esteem, how is it possible that the self-esteem of black children is not lower than of white children? The flaw in this reasoning, we suggest, is *the assumption that (at least among children), blacks are using whites as their comparison reference group.* Social comparisons do affect self-esteem, and do so for blacks and whites alike. But overwhelmingly, we believe, the black child compares himself with other blacks, not with whites. And he does so on the basis of the actual structure of the environment in which he predominantly lives.

Data demonstrating the deleterious self-esteem consequences of dissonant academic, socioeconomic, and family contexts have already been presented in chapters 2 and 4. Assuming that children and adolescents tend to use those in their immediate environments as their comparison reference groups, the question we wish to raise here is: to what extent *are* black children actually in such dissonant contexts and thus likely to use white children as their comparison reference groups?

Consider school marks in the Baltimore study. As indicated in our discussion of contextual dissonance (Chapter 4), black pupils in white schools earn better marks than black students attending black schools; hence, it would follow that their self-esteem should be higher. In fact, however, it is lower.[14] The reason is that the black children in predominantly white schools are comparing themselves with white pupils in their schools, where, we showed earlier, they suffer by comparison, rather than with other black pupils in black schools, where they would gain by comparison.[15]

But, one might ask, shouldn't the academic comparisons with whites in desegregated schools reduce the average self-esteem level of black children in the sample as a whole? The answer is that it does, but not by much, because the vast majority of black children in this sample live in predominantly black environments. Only 3 percent live in predominantly white neighborhoods and 12 percent attend predominantly white schools.[16] Most therefore compare themselves with other black children in their own environments; hence, the overall self-esteem level is little affected.

Or consider the issue of black poverty. As in the rest of America, the Baltimore black children, on the average, are much more heavily concentrated in the lower classes than the white. The interesting point is that, *within their schools and neighborhoods,* poor white children, not black children, are more likely to be economically inferior, that is, to suffer from economic comparisons. It may be recalled from chapter 4 that nearly half of the lowest class (Hollingshead Class V) blacks attend schools in which their socioeconomic status is about average compared with none of the Class V whites, all of whom are socioeconomically inferior in these school contexts. Among Class IV (working class) blacks, one-fourth are *above average* for their school and one-tenth below average. For the Class IV whites, none is above average and one-third are below average. Though blacks are substantially poorer than whites, it is the whites who fare more poorly in socioeconomic comparisons with their schoolmates and whose self-esteem may be deleteriously affected.

Third, consider family structure. Black children are much more likely than white ones to come from the most stigmatized family structures—separated, abandoned, or never-married families (Farley and Hermalin, 1971: 8). And, as a matter of fact, if black children were comparing themselves with whites, their self-esteem would suffer. As we noted in the earlier discussion of contextual dissonace (Chapter 4), in predominantly white schools, the self-esteem of black children from separated or never-married families is considerably below that of black children from intact families, but in predominantly black schools, no self-esteem differences appear. The reason is that black children in Baltimore from separated or never-married families overwhelmingly attend schools in which some kind of family rupture is common. Comparing themselves with those around them, their self-esteem suffers no damage. The damage is inflicted on the self-esteem of black children from stigmatized families [17] who attend predominantly white schools, where they experience unfavorable comparisons. These particular children, however, *constitute only a small proportion of all the black children in this type of family* (*11 percent*) *and an even smaller proportion* (*3 percent*) *of all* black children in the entire Baltimore sample. The upshot is that there is little *overall* difference in the self-esteem of black children from separated or never-married families and black children from other family origins.

Finally, let us consider the issue of skin color. Despite protestations to the contrary, the evidence is clear in showing that black children in the sample implicitly consider light skin more attractive than dark. When black children are asked whether lighter or darker skin color is nicer or whether both are the same, they usually reply that both are the same. But when asked which person has the nicest skin color in his family, and whether that person's color is lighter or darker than his own, 74 percent say "lighter," 6 percent "darker," and 20 percent "the same." Furthermore, among black interviewers (but not among white interviewers), black children described as having light skin are more likely to be described by the interviewers as "very good-looking." By virtue of the darker skin color which they themselves evidently consider less attractive, then—and numerous theorists have interpreted this fact as generating self-hatred [18]—blacks would be expected to have low self-esteem.

But what are the facts of the case? First, it should be stressed that looks are important for the self-esteem of both black and white children. For both races, the lower the child's opinion of his looks, the lower his global self-esteem. Second, the darker the child's skin (as judged by the interviewer), the less likely he is to consider *himself* good-looking. But do black children, having darker skin than whites but valuing lighter skin, therefore consider themselves less good-looking? Emphatically not. At each age level, according to the Baltimore data, they are *more* satisfied with their looks than white children are with theirs. The reason is obvious. The black child compares his skin color with that of blacks, not whites, thus emerging with a distribution of satisfaction with looks similar to that of any other group. Lighter-skinned blacks are more likely to consider themselves very good-looking than darker-skinned blacks, not to consider themselves less good-looking than whites do.

## Discussion

Like the reflected appraisals principle, the social comparison principle is entirely sound, receives support from empirical data, and has obvious self-esteem implications. If it fails to yield correct empirical predictions in this case, it is not because the principle is wrong but because the wrong referent other has been selected. Most social scientists appear to have operated on the implicit assumption that minority children use the dominant majority as their comparison reference group, an assumption we have serious cause to question.

The error, to be sure, is understandable. The social scientist has ample reason to expect poverty, poor school performance, stigmatized family origins, and disvalued physical type to damage self-esteem. And, as he examines the objective statistical data based on carefully selected samples, he sees that blacks compare unfavorably with whites in all these respects. His conclusion that blacks must have lower self-esteem is thus entirely logical.

His reasoning is, in fact, sound in those circumstances in which blacks actually do use whites as their comparison reference groups. The Baltimore data have clearly indicated that those black children who interact chiefly with whites (desegregated contexts), and who therefore compare unfavorably in terms of socioeconomic status, academic performance, and normatively desirable family structure, *do* apparently experience reduced self-esteem. But most black children, the data also show, interact primarily with other black children, and tend to compare themselves chiefly with those with whom they directly interact. No one could visit a black school—100% black in Baltimore in 1968—walk around the neighborhood, watch the children at play, without feeling that this was a self-contained, almost hermetically sealed, black world. This is the world in which life is lived; the outside world is relatively distant and unreal.

In sum, a vast amount of popular and scholarly literature has appeared concluding that minority group members have lower self-esteem because (1) they are looked down upon by others (the direct reflections principle); (2) they look down on themselves by virtue of internalizing the negative attitudes of society toward their group (the generalized other); or (3) they compare unfavorably in important evaluated respects with the prestigious majority (the social comparison principle). All these principles are unquestionably sound and represent important determinants of self-esteem. The reason these principles may generate erroneous conclusions regarding minority status and self-esteem is that they may be applied in the face of certain contrary-to-fact assumptions: the assumption of full awareness; the assumption of agreement; the assumption of personal relevance; the assumption of interpersonal significance; and the assumption of majority referent others. Since these assumptions are largely without foundation, it is understandable that the various theoretical predictions, though based on sound principles, should go astray.

## NOTES

1. As Cartwright (1950: 440) observed: "The groups to which a person belongs serve as primary determiners of his self-esteem. To a considerable extent, *personal* feelings of worth depend on the social evaluation of the *groups* with which the person is identified. Self-hatred and feelings of worthlessness tend to arise from membership in underprivileged or outcast groups."

2. Other writers, both popular and scholarly, have often simply assumed low self-esteem among blacks as a self-evident, fundamental, and irreducible datum, and proceeded from there (Erikson, 1966; Ausubel, 1958; Pettigrew, 1964; Kvaraceus, *et al.*, 1965). The discordant voices in the nearly unanimous chorus of opinion (such as McCarthy and Yancey, 1971; Baughman, 1971; Luck and Heiss, 1972; Wylie, 1978) have been relatively few.

3. These data are not directly comparable to those of earlier decades, which were based on different methods and sampling procedures (e.g., Clark and Clark, 1958; Goodman, 1952; Brody, 1963). It is not certain whether the differences in the pre- and post-sixties studies are due to differences in methods or changes in self-concepts.

4. This analysis is restricted to pupils whose parents had the same background.

5. Luck and Heiss' (1972) sample of adult males showed similar results.

6. These measures, discussed in earlier chapters, are described in Appendix A.

7. Although the number of black Catholics in the sample is small, Hunt and Hunt (1975) also found that black Catholics had somewhat higher self-esteem than black Protestants.

8. These studies represent only a small sample of a vast literature which has received exhaustive coverage and probing examination by Wylie (1978). The conclusion that there is no compelling evidence of lower self-esteem among minority group children is based not simply on the studies cited above, but also on several literature reviews (Wylie, 1978; Christmas, 1973; Guzman, 1976).

9. In a well-documented report, Milner (1953) also indicated that full awareness of the status of his group does not usually occur to the black child until early adolescence.

10. Even accepting the same system of prejudice, then, it is possible for the high status group to enhance its self-esteem by emphasizing its difference from the low, whereas the low status group enhances its self-esteem by emphasizing its similarity to the high. For example, Sherif (1962: 809) presented a sample of adolescents in the Southwest with a list of ethnic groups and asked the children to indicate which groups were "like" them. The native born whites denied that the blacks were like them, but the blacks affirmed that they were like the native born whites.

11. Erik Erikson (1966: 155) contends that "the individual belonging to an oppressed and exploited minority, which is aware of the dominant cultural ideals but prevented from emulating them, is apt to fuse the negative images held up to him by the dominant majority with his own negative identity . . . There is ample evidence of 'inferiority' feelings and of morbid self-hate in all minority groups." Unfortunately, he neglects to cite any of this "ample" evidence.

12. It has been argued (Clark, 1965) that black children's teachers fundamentally have low respect for their pupils and have little expectation that they can and will learn. Whether or not this is true, we are unable to say. But we can report that, in the Baltimore study, the black children *do not perceive it* that way; they are just as likely as the white children to say that their teachers think well of them.

13. Though the darker child is less likely than the lighter to consider himself very "good-looking," he does not necessarily have lower global self-esteem. (See Rosenberg and Simmons, 1972.) Feelings about one's looks, however, are definitely related to global self-esteem for both races.

14. This finding in Baltimore is supported by a substantial body of other research (see St. John, 1975, and Simmons *et al.*, 1977).

15. These data thus support the prediction of McCarthy and Yancey (1971: 665) that "the degree to which blacks employ whites as significant others ought to vary directly with informal contact with whites. Though such contacts may be selective, it might be hypothesized that variation in contact is negatively related to self-esteem."

16. Baltimore is far from unique in this regard. Taeuber and Taeuber (1965), in a study of American cities, including Northern ones, between 1950 and 1960, showed that less than one percent of the neighborhoods in many Northern cities were stable, racially integrated neighborhoods. Similarly, Kenneth Clark (1965: Ch. 3) showed the racial residential segregation in major American cities to be massive.

17. The self-esteem of white children (in predominantly white schools) from such families also suffers severely.

18. See Grier and Cobbs' (1968) discussion of black women's hatred of their skin color.

# 7

# Group Rejection
# and Self-Rejection

THE PREVIOUS CHAPTER dealt with the impact on self-esteem of *society's* negative attitude toward the individual's group. The present chapter considers the effect on self-esteem of the *individual's own* negative attitude toward his group. This theme is reflected in the various discussions of "self-hatred among Jews" or "black self-hatred" which have appeared over the years. In essence, the reasoning has been that, since the group is an important part of the self-concept, rejection of the group thereby constitutes rejection of the self.

One of the most important early statements of this position appeared in a series of articles by Kurt Lewin (1948) on the theme of "self-hatred among Jews," which has since strongly influenced social psychological thought on minority group membership. In the period of the European ghetto, Lewin pointed out, the Jews were confined to special sections of the city, were restricted in their freedom of occupational choice, and were easily recognized by their yellow badge, special garb, and other overt signs. Such stigmatization and restriction generated a good deal of tension, but at least there was no problem of self-identification. The Jew accepted himself as a Jew and identified as a Jew. With the Emancipation, however, Jews were able to expand into the broader society and were no longer easily distinguished from the general population. But social boundaries in the form of prejudice remained. Seeking to achieve acceptance into the higher status group, the Jew's ties with his religious group weakened; at the same time, prejudice prevented him from achieving complete acceptance by the outside group. Socially identified as a Jew, he experienced ambivalent self-identification and self-hatred.

Speaking of mid-twentieth century America, Lewin argued that "Jewish self-hatred is a phenomenon that has its parallel in many underprivileged groups. One of the better known and more extreme forms of self-hatred can be found among American Negroes . . . An element of self-hatred which is less strong but clearly distinguishable may also be found among the second generation of Greek, Italian, Polish, and other immigrants to this country" (Lewin, 1948: 189).

When one examines the term "self-hatred" more carefully, it becomes apparent that two different meanings are involved. The first is hatred or rejection of one's self as a person, that is, *individual* self-hatred or low self-esteem. Lewin's primary interest, on the other hand, was in what might be called "group self-hatred," i.e., rejection of, or disidentification with, one's *group* (for example, religion, race, nationality). Most writers have assumed that, since one's group is an important part of one's self, rejection or hatred of that group reflects rejection or hatred of oneself. According to Proshansky and Newton (1968): "The Negro who feels disdain or hatred for his own racial group is expressing—at some level of awareness—disdain or hatred for himself." The aim of this chapter is to examine the connection between the individual's feelings toward his group and his feelings toward himself.

## Group Rejection and Self-Rejection

When one considers the term "group rejection" carefully, it is evident that several different nuances of meaning are associated with it. We have attempted to capture some of these distinctions in the

Baltimore study. One of these nuances refers to *lack of pride* in one's group. As a crude indicator of this concept, we asked our Baltimore respondents: "How proud are you of being Negro?", "How proud are you of your religion?", "How proud are you of your school?", and so forth.

A second aspect of group rejection deals with introjection. This somewhat elusive idea refers to the degree to which the group is experienced as an integral and inseparable part of the self. Introjection, as we pointed out in chapter 1, refers to the "adoption of externals (persons or objects) into the self, so as to have a sense of oneness with them and to feel personally affected by what happens to them." For the group identifier, the distinction between *me* and *my group* is unclear; the fate of the group is experienced as the fate of the self. In Baltimore, group identification was chiefly indexed by the question: "If someone said something bad about the Negro or colored race; your religion; your school; etc. would you feel almost as if they had said something bad about you?" [1]

A third dimension of group rejection is importance. Does the particular self-concept component (race, religion) rank low or high in the individual's hierarchy of self-values? Although the individual may not lack pride in his group or deny that it is part of him, he may still consider it personally unimportant. Members of third or fourth generation immigrant groups, nominal members of religious groups, and graduates of certain high schools might be examples. In Baltimore, this aspect of group rejection was indexed by the question: "Is being Negro or colored; your religion; your school; etc. very important to you, pretty important, or not very important to you?"

These three indicators of group pride, introjection, and importance regarding "your religion" were combined to form a score of "religious group rejection or attachment"; the same three indicators were combined with three others to form a score of "racial group rejection or attachment" (described in chapter 4).[2]

We can now examine the assumption, implicit in the work of Lewin, Clark and Clark, and most other writers in this area, that rejection of one's group involves, virtually by definition and hence as an empirical necessity, rejection of oneself. Turning first to *race*, Table 7–1–A examines the relationship between the black child's rejection of his race and his global self-esteem. It is apparent that the two are virtually unrelated. The global self-esteem of those black respondents whose racial attachment is relatively weak differs little from that of those with strong racial introjection, pride, and importance.

Turning to *religious* group rejection (Table 7–1–B), the results are virtually the same. Contrary to theoretical expectation, rejection of one's religious group does not appear to be associated with any propensity to reject oneself as an individual.

**TABLE 7-1.**

*Racial and Religious Group Attachment
and Self-Esteem*

| | A. | | |
|---|---|---|---|
| | Racial Group Attachment (Blacks) | | |
| | Strong | Medium | Weak |
| *Self-Esteem* | | | |
| Low | 19% | 16% | 20% |
| Medium | 35 | 39 | 34 |
| High | 46 | 45 | 46 |
| N = 100% | (327) | (158) | (728) |
| | B. | | |
| | Religious Group Attachment | | |
| | Strong | Medium | Weak |
| *Self-Esteem* | | | |
| Low | 25% | 27% | 29% |
| Medium | 31 | 33 | 30 |
| High | 44 | 40 | 41 |
| N = 100% | (613) | (414) | (245) |

There is little empirical research on the relationship of group self-hatred to individual self-hatred, perhaps because investigators have considered it pointless to examine what appeared to be a virtual tautology. A modest amount of additional evidence is, however, available. One finding appears in the New York State study of high school juniors and seniors. The only indicator of group attachment available in that study was the question: "How important is your religious group to you?" Table 7–2 indicates that the adolescent's feeling that his religious group is or is not important to him bears only a slight relation to his global self-esteem. Furthermore, whatever small relationship there is appears among Catholics and Protestants,

**TABLE 7-2.**

*Importance of Religious Group and
Self-Esteem (New York State)*

|  | Importance of Religious Group | | |
|---|---|---|---|
|  | Very | Fairly | Not |
| *Self-Esteem* |  |  |  |
| High | 46% | 48% | 42% |
| Medium | 26 | 24 | 24 |
| Low | 28 | 28 | 34 |
| N = 100% | (1244) | (841) | (206) |

not among Jews, the group to which the theory most appropriately applies.

In the previous chapter, the issue of why *society's* rejection of our group is apparently unconnected with our self-esteem was examined from the viewpoint of reflected appraisals and social comparison processes. In this chapter, dealing with why *our own* rejection of our group is unrelated to self-esteem, the principle of psychological centrality is invoked as the major explanatory factor.

## The Psychological Centrality Assumption

The self-concept, we stressed in chapter 1, is an extremely complex structure, consisting of a very large number of social identity elements, traits, physical characteristics, abilities, interests, ego-extensions, and other components. In centering attention on a particular social identity element, (such as race or religion) it is easy to overlook the fact that his identity element is not the sole psychologically central component of the self-concept. To the black, Jew or Mexican-American, there is more to the self-concept than being black, Jewish or Mexican-American. A person is not only black but also good-looking or popular; not only Jewish, but also musically talented and athletically adept; not only Mexican-American, but also rich and respected. Hence, even if an individual does experience "group self-hatred," if other self-concept components, individually or collectively, weigh more heavily in his scale of values, then such negative feelings toward his group may have only a minor impact on his feelings toward himself.

## COMPARISON WITH OTHER SELF-CONCEPT COMPONENTS

The child's race or religion, we shall see, are important to him. The reason they are not decisive for self-esteem, however, is that they (1) may be matched or overshadowed by certain ego-extensions, (2) may be less important than certain dispositions, and (3) may fall short of the importance of individual achievements.

*Comparison with Other Ego-Extensions.* The individual's self-feeling as pointed out in chapter 1, extends far beyond the boundaries of his skin to incorporate all objects eliciting the distinctive feeling of "me" or "mine," for example, mother, father, siblings, home, school, clothes, friends. These ego-extensions—elements technically external to the self—are nevertheless experienced as part of it. In order to see where race and religion fit into the total self-concept structure, we have compared these identity elements with other ego-extensions in terms of the three dimensions of attachment or rejection—introjection, pride, and importance—mentioned above. In Table 7–3, Column A, the Baltimore children's ego-extensions are ordered in terms of the frequency with which respondents averred that an attack on the ego-extension would be felt as an attack upon themselves; this indicator of introjection is probably the surest sign of "identification." With regard to "importance" and "pride," (Columns B and C), we were able to ask about only six ego-extensions. Although the ego-extensions in the three lists are not identical, they are comparable and their rank orders, computed by a Spearman Rank Order Correlation, are very similar. (Rhos are as follows: Columns A and B = .90; Columns A and C = .70; Columns B and C = 1.00).

The essential point to be drawn from Table 7–3 is that, although race and religion are widely felt to be an integral part of the self, mother, father, and siblings are even more generally accepted as ego-extensions; such other objects as clothes, school work, friends, father's job, and school are also chosen by more than half the sample. Furthermore, Columns B and C show that, although it is true that race and religion are considered "very important" by most children and that they are "very proud" of these group affiliations, a number of other ego-extensions also appear to involve the self deeply. Thus, while race and religion are certainly important, they are not of

**TABLE 7-3.**

*Identification with Potential Ego-Extensions*

| | A | B | C |
|---|---|---|---|
| | Attack on ego-extension felt as attack on self | Ego-extension "very important" | "Very proud" of ego-extension |
| | % | % | % |
| *Potential ego-extension* | | | |
| Mother | 88.5 | | |
| Father | 82.6 | | |
| Family | | 86.2 | |
| Sibs | 80.0 | | |
| Negroes (black children only) | 77.6 | 67.0 | 79.2 |
| Religion | 74.4 | 64.8 | 69.5 |
| Home | | | 65.2 |
| Clothes | 71.2 | | |
| Schoolwork | 66.6 | | |
| Friends | 61.9 | 34.3 | 39.5 |
| Father's job | 59.5 | | 60.6 |
| School | 50.0 | 57.1 | 48.4 |
| Toys | 31.5 | | |
| Neighborhood | | 24.8 | 32.0 |
| President of U.S. | 27.2 | | |
| Governor of State | 24.2 | | |

exclusive, and may not even be of paramount, importance compared to other groups, individuals, or objects with which the individual may identify.

*Comparison with Dispositions.* Not only are the individual's dispositions numerous, but they are often experienced as touching the very core of the self. In certain cases, our social identity elements are felt to be the external, superficial aspects of our selves while our traits or personal attributes may be felt to be more of what we truly are; if so, the psychological centrality principle would suggest that global feelings of worth would depend more heavily on the evaluation of traits.

One way to test this assumption is to see whether the individual's pride or shame in certain traits (intelligence, appearance, social skill) are more or less closely related to his global self-esteem than his pride or shame in certain aspects of social identity (race or religion). If one looks at Table 7–4, it is immediately apparent that the relationship between degree of pride in one's race, or pride in one's religion, is much more weakly associated with the youngster's global self-esteem than his assessment of how smart, good-looking or easy to get along with he is. The relationships of self-esteem to assessment of the three dispositions are gamma = .2564, .2629, and .1868, whereas the relationships of self-esteem to pride in the two social identity elements are gamma = .0823 and .0872. Insofar as all of these parts contribute to the whole, the contribution of dispositions would appear to be greater (though the data are only suggestive of this conclusion).[3]

This is not to imply that all traits are more central to the individual's feeling of personal worth than all social identity elements. The only point is that, given the enormous number of traits available, and the prominence of some of these, it is unlikely that feelings about one's group will exercise a decisive influence on global self-esteem.

### TABLE 7-4.
*Relationship of Self-Esteem to (A) Three Traits and (B) Two Social Identity Elements*

|  | Gamma | N |
| --- | --- | --- |
| *Relationship of self-esteem to . . .* |  |  |
| *Traits* |  |  |
| How smart are you? | *.2564 | (1831) |
| How good-looking are you? | *.2629 | (1735) |
| How easy are you to get along with? | *.1868 | (1811) |
| *Social Identity Elements* |  |  |
| How proud of your race? (blacks) | .0823 | (1073) |
| How proud of your religion? | .0872 | (1649) |

*See footnote 7, chapter 3, p. 98.

*Comparison with Achievements.* In chapter 5, we pointed out that one reason why social class bore so little relationship to global self-esteem among children was that, for the child, it was an ascribed rather than an achieved status. In those areas of achievement actually available to the child (school marks and election as club officer), the data were consistent with the self-attribution principle in showing an association between overt achievement and self-attitudes. Social class represented the outcome of the father's, not the child's, efforts.

But race, religion, or ethnicity represent the outcome of *no one's* efforts, neither father nor child; they are totally ascribed. If children were indoctrinated with the normative belief that the worth of a person is based on what he has actually achieved rather than on the accident of birth, then, following the self-attribution principle, their ascribed statuses might have relatively little importance for their self-esteem; their true worth would rest on the firmer foundations of personal achievement.

Let us therefore consider whether the two areas of childhood achievement discussed in chapter 5—school marks and election to club office—are more closely related to self-esteem than the individual's level of pride in the ascribed statuses of race and religion. (Because elementary schools do not have clubs, we shall limit this discussion to secondary school pupils). Among this group, the data show, the relationship of self-esteem to school marks and to election as a club president are gamma = .1298 and .1281, whereas the relationship of self-esteem to race pride and religious pride is gamma = .0403 and .0184. Since the "club president" question is based on the comparatively small number of pupils who belong to school clubs, however, the results may be due to statistical chance.

**TABLE 7-5.**

*Relationship of Some Ascribed and Achieved*
*Statuses to Self-Esteem, for Secondary School Pupils*

|  | Gamma | N |
|---|---|---|
| *Relationship of self-esteem to . . .* | | |
| *Ascribed statuses* | | |
| Proud of race (blacks) | .0403 | (437) |
| Proud of religion | .0184 | (784) |
| *Achieved statuses* | | |
| Marks in school | *.1298 | (743) |
| President in club (for club members) | .1281 | (247) |

*See footnote 7, chapter 3, p. 98.

It is thus evident that the youngster's social and academic achievements—which represent the major realms of achievement at this age—are certainly associated with global self-esteem, either as cause or effect. On the other hand, since racial and religious statuses can in no sense be interpreted by him as products of his own efforts or expressions of his essential characteristics, they understandably bear little relationship to his global feeling of worth.

In sum, although much has been written on the general theme of group "self-hatred," most of it has overlooked the critical issue of the relationship of the parts to the whole. The self-concept is a structure, and an intricate one at that. The individual's race is only one element among many which bears upon his global self-esteem, and it is far from paramount to his concerns.

## Group Pride and Group Introjection

When Lewin (1948), Kardiner and Ovesey (1951), Proshansky and Newton (1968), and others speak of group "self-hatred," they are essentially referring to a feeling of *lack of pride* in one's group; the individual is assumed to share the low regard in which his group is held by the wider society. It is essentially for this reason that a number of recent sociopolitical movements have focused on enhancing group pride (black pride, yellow pride, pride in *la raza*) as a means of raising the self-respect of its members.

Although group pride and group introjection are obviously related, it is useful to highlight the distinction between them. Whether all the respondents who claimed to be "very proud" of their group did in fact feel such strong group pride is uncertain, but it seems reasonable to think that the small number who said they were "not proud" of their race or religion probably did hold such negative feelings. Presumably, it is this group that writers have in mind when they speak of "group self-hatred." Lack of introjection, on the other hand, means that the group is not experienced as an integral part of the self, and that the individual does not "feel personally affected by what happens" to the group.

Given this distinction between low group pride and lack of introjection, we would advance the following curious proposition: among people with low group pride, lack of introjection, rather than injuring self-esteem, may actually protect it. The principle is not obscure. Take a student attending Community Junior College who knows that his school has low prestige and, indeed, agrees that this poor reputation is justified; this attitude certainly expresses "group self-hatred" or low pride in his group. But assume that his sense of personal worth is totally separated from the school's reputation ("I just go there"); in this case, his self-esteem would be unaffected. In other words, it would be group introjection, not its absence, that would damage self-esteem.

We attempted to test this reasoning in the Baltimore study. For example, consider those black children with "racial self-hatred," that is, the small proportion who say they are "not proud" of their race. Table 7-6 indicates that if their racial introjection is strong (attack on race is experienced as attack on self), then 46 percent have low self-esteem; but if they do not introject their race, then only 23 percent have low self-esteem. For those with racial self-hatred, lack of introjection appears to *protect* individual self-esteem.

Similar results appear with respect to religious group introjection. Once again, we confine our attention to those with "religious self-hatred," that is, those "not proud" of their religion. Among those who introject the group, 55 percent have low self-esteem, whereas among those who do not, only 30 percent have low self-esteem.

**TABLE 7-6.**

*Self-Esteem and Group Identification Among Pupils*
*Expressing Low Group Pride*

|  | Black students with *low* pride in race who: | | Students with *low* pride in religion who: | |
|---|---|---|---|---|
|  | Identify | Disidentify | Identify | Disidentify |
| Low Self-Esteem | 46% | 23% | 56% | 30% |
| N = 100% | (11) | (22) | (18) | (23) |

Because so few children actually manifest low racial and religious pride, these differences are not statistically significant; plainly, the conclusions cannot be accepted without further research. Nevertheless, they strongly call into question the widespread assumption in literature that the individual who rejects his group is necessarily rejecting himself. Under conditions of actual group self-hatred—feelings of low pride in, or respect for, one's group—such failure to introject the group may *protect* one's feeling of self-worth while introjection may *harm* it.

Some important causal implications inhere in these observations. Theoretical reasoning suggests that if a group is derogated in the society, rejection of the group will eventuate in low personal self-esteem (Proshansky and Newton, 1968). In light of the foregoing, however, it is possible that *group introjection may be as much a consequence as a cause of self-esteem*. If, as suggested in chapter 2, a fundamental human motive is the defense and enhancement of the self, then it is possible that introjection may be motivated by a desire to protect and enhance one's feeling of self-worth. In other words, if the individual feels pride in his group, then we would expect him to introject it (consider it an integral part of himself), thus enhancing his self-esteem; but if he does not feel proud of his group (perhaps because he internalizes society's negative attitudes toward his group), then he would be motivated not to introject it, *thereby also protecting his self-esteem*. Attitudes toward one's group may be a cause of self-esteem, but the motive to protect self-esteem may be in part responsible for attitudes toward one's group.

The data in Table 7–7 are consistent with this interpretation. There we see that respondents tend to introject groups of which they are proud and fail to introject groups of which they are not. Both outcomes protect self-esteem. It is only those who introject groups of which they are *not* proud who suffer blows to their self-worth. Such people probably represent the archetypes for the theoretical discussions of self-hatred among Jews, blacks, and other minority groups. According to our data, they are few in number, a finding consistent with Guttentag's (1970) literature review showing that pride in, and identification with, ethnic groups—whatever their prestige level in the society—remains high today.

The outcome for self-esteem is thus a happy one. We tend to consider as an integral part of ourselves those membership groups we admire but as separate from ourselves those membership groups

**TABLE 7-7.**

*Pride in Group and Introjection of Group*

| "If someone said something bad about [group], would you almost feel as if they had said something bad about you?" | "How proud are you of your [group]?" | | |
|---|---|---|---|
| | Very proud | Pretty proud | Not very proud |
| *Would feel bad if *race* insulted | 82% | 72% | 31% |
| N = 100% | (823) | (170) | (35) |
| *Would feel bad if *religion* insulted | 83% | 59% | 44% |
| N = 100% | (941) | (374) | (41) |

*See footnote 7, chapter 3, p. 98.

we disdain. Even in a society characterized by widespread prejudice and unequal group evaluation, individuals can avail themselves of mechanisms to protect and enhance their self-regard.

This is not to suggest that group rejection may not have *other* noxious consequences for the individual. The pain and bitterness of isolation from one's group is a recurrent theme in fiction and social science literature. The feeling of belonging, the experience of mutual acceptance, and the sense of security that comes from unity with a group is powerful indeed (Fromm, 1941; Maslow, 1954; Durkheim, 1951). In discussions of group disidentification or self-hatred, however, there is one point which is widely overlooked, namely, that *it is not the group that rejects the individual but the individual who rejects the group.* There is all the psychological difference in the world between these two situations. The Jew who wishes to identify with his group is welcomed by his coreligionists with open arms. Self-esteem is affected by rejection *by* the group, not *of* the group. If, on the basis of personal choice, the individual refuses to identify with other members of his minority group, why should this make *him* feel inferior? He may, on the contrary, consider himself better than other members of his group. One can see why the individual's rejection of his group may contribute to insecurity by producing feelings of isolation but not why it should contribute to feelings of personal worthlessness.

## Summary

Although the idea that rejection of one's group results in rejection of one's self is usually discussed in the context of racial, religious, or ethnic group membership, its theoretical implications are broader, for similar issues surround any other negative group or status, for example, homosexual, prostitute, delinquent, embezzler. The general question is: does disdain for the particular social identity element necessarily result in disdain for the self as a whole?

The answer to this question, we have suggested, requires us to view the self-concept as a total structure and to take account of the precise relation of the parts to the whole. The self-concept consists of a multiplicity of components, including dispositions and ego-extensions, having different priority, centrality, and weight in determining the individual's global self-evaluation. These components are selectively perceived, evaluated, and ordered. Our analysis of race and religion suggests that these social identity elements are less likely to be central to the self than certain other ego-extensions and to contribute less to self-esteem than some traits or areas of achievement.

We have also observed that group rejection or disidentification is a more complex idea than it initially appears, for it involves the issues of pride, introjection, and importance, and that these do not always coincide. The result is that even if people do share society's low regard for their group, this attitude need not deleteriously affect their global self-esteem if the group is not introjected, that is, experienced as an integral, inseparable part of the self. Needless to say, such group identification is not purely accidental. People are motivated to protect their self-esteem, and they consequently tend to introject those groups of which they are proud but not those of which they are not.

## NOTES

1. Herman (1970: 67) also used this manifestation of introjection as an important indicator of group identification or rejection. He asked students in Israel: "When an important overseas journal insults the Jewish people, do you feel as if it was insulting you?" (In this sample, 87 percent replied "always" or "often").

2. The indicators of "religious group disidentification" and of "racial group disidentification" are presented in Appendix B-3.

3. The relative importance of variables, of course, is not adequately assessed by a comparison of zero-order associations; hence, the data are only suggestive. For discussions of this issue, see Blalock (1961) and Rosenberg (1968).

# PART III

# Self-Concept Development

THE VIEW of the self-concept advanced in this work—as the totality of the individual's thoughts and feelings with reference to himself as an object—is plainly that of an intricate structure. But this self-concept is a product of development; it is not there at birth, it does not emerge full-blown overnight, and, once established, it does not remain unchanging for eternity. The self-concept is built up gradually out of social experience in the course of maturation. Although sensitive and insightful observations on the emergence of the consciousness of self were presented by Baldwin (1897) and Cooley (1912) in the late nineteenth and early twentieth centuries, an adequate understanding of the development of the self-concept, in all its intricacy and complexity, still awaits us. Conspicuously lacking is information on development *after* early childhood. There is an influential school of thought which holds that personality (and, presumably, the self-concept as an important personality component) is fundamentally formed in the early years and that subsequent developments are symbolic recapitulations or elaborations of the themes laid down in the earlier years. Although we agree that the early years are important, we do not concur with the view that they are of exclusive importance, least of all with regard to the self-concept. The self-concept is a product of relatively late development, and, as we shall see, many fundamental changes take place in middle childhood right through adolescence. In time, of course, the self-concept develops some degree of stability, but it is never fixed in concrete; changes, of a gradual or sudden sort, continue throughout the life span.

Chapters 8 through 10, therefore, deal with the relatively unexplored area of self-concept change and development from the period of middle childhood (about 8–9 years old) through late adolescence (about 18–19). Chapter 8 takes up an area of self-concept structure mentioned, but not elaborated, earlier, namely, the exteriority or interiority of the self-concept. Chapter 9 turns to the issue of self-concept disturbance. Are there periods of life when the self-concept, in whole or in part, is in a state of turmoil, characterized by instability, self-doubts, high self-consciousness, and other stressful states? Finally, chapter 10 returns to the issue (originally discussed in chapter 3) of interpersonal significance, but viewed here from a developmental perspective. As noted earlier, children have different levels of faith in other people's knowledge of what they are like. The question considered in chapter 10 is whether this locus of self-knowledge changes as children grow older and what some of the consequences of this shift may be. As we shall see, powerful changes in how youngsters *think* about themselves (chapter 8), how they *feel* about themselves (chapter 9), and where they feel the *truth* about themselves lies (chapter 10) occur in the course of development.

# 8

# Exterior and Interior Self-Concept Components

THE PROMINENCE of a self-concept component, we have observed repeatedly in previous chapters, importantly influences its impact on the individual's global self-feelings. But there is another aspect of psychological centrality to be considered, defined not in terms of the emotional involvement in the specific components, but in terms of their exteriority or interiority. Which of these predominate in the total self-concept structure, and how does such predominance shift as children grow older? It is to these questions that the present chapter is addressed.

Each of us, it may be suggested, has two selves: an overt or revealed self and a covert or concealed self. The overt self represents those aspects of the self which are generally public and visible, such as our physical, demographic, or behavioral characteristics. These might be said to reflect the social exterior of the individual. But parallel with this social exterior is a psychological interior, a private world of thoughts, feelings, and wishes which is either totally or relatively inaccessible to the world outside. This distinction, we shall argue, is crucial to understanding the difference between the self-concepts of younger and older children. Furthermore, it is directly traceable to the most fundamental of social processes—human communication.

## Development of Exterior and Interior
## Self-Concept Components

Does the tendency to conceptualize the self in these terms change as children grow older? This question was investigated in the Baltimore school sample by presenting these subjects, aged 8–19, with a number of open-ended questions dealing with diverse aspects of the self and observing how they spontaneously structured their responses. Although the answers naturally differed depending on the particular content of each question, a general tendency to conceptualize the self in terms of a social exterior or a psychological interior was evident.

At various points in the study, our subjects were asked the following six questions or sequences of questions:

1) Locus of self-knowledge. "Let me ask you this: Who do you feel really understands you best? I mean, who knows best what you really feel and think deep down inside?" As noted in the earlier discussion of significant others, this question was followed by several choices, including one's mother, father, friends, brothers and sisters, and teachers. Following responses to these structured questions, we inquired: "Let me ask about the person who knows you best. What does this person know about you that other people don't?"

2) Points of pride. "Most people have certain things that are better about them than other things. How about you? Could you tell me what things are really best about you, are the best part of you?"

3) Points of shame. "Most people have certain things that are not as good about them. Do you have any weak points, that is, any things not as good about you?"

4) Sense of distinctiveness. "How different are you from most other kids you know?" (If "very" or "fairly"): "In what ways are you different?"

5) Sense of commonality. "How much are you the same as other kids you know?" (If "very" or "fairly"): "In what ways are you the same?"

6) Future self. "Now I would like to ask you what kind of person you would like to be when you are an adult. . . . What kind of personality would you like to have when you grow up? I mean, what kind of person would you like to be?"

Given the general, open-ended nature of these questions, we would naturally expect younger children to be more likely to give vague, ambiguous, or no responses; and this is indeed the case. For purposes of comparability, then, we will include in our analysis only those subjects who gave meaningful substantive answers, recognizing, of course, that these respondents do not perfectly represent their age cohorts. Despite its limitations, this procedure seemed preferable to attempting to read into the child's mind what he really thought when he answered "nothing," "everything," "like I am now," "average," "don't know," and similar responses.

WHAT OTHERS KNOW ABOUT US

When asked their views on the locus of self-knowledge, that is, what the person who knows them best knows about them that other people do not, how do children of different ages respond? The most striking finding is the degree to which the older child, relative to the younger, is likely to answer in terms of some aspect of his psychological interior. In response to this particular question, the majority of older children cite either (1) general thoughts and feelings; (2) specific interpersonal feelings, or (3) private wishes, desires, or aspirations.

First, adolescents are much more likely than younger children to answer the question in terms of general *inner thoughts or feelings*. What is of special interest in these replies is that the psychological interior is conceptualized as a broad melange of phenomenal states, knowledge of which is vouchsafed to only a select few. The special knowledge, then, is not necessarily a specific attitude, idea, feeling, or secret but a miscellany of affective and cognitive components, a realm of thoughts and feelings. More specifically, adolescents tend to say that those who know them best know "what I'm really like," "the real me," "how I feel deep down inside when I'm hurt," "what I really mean." Particularly prominent are references to emotional states. The knowledgeable other understands our deeper feelings: "they know if I'm happy or not," "my feelings," "that I worry a lot," "can tell when I'm sad or when something is wrong," "know my feelings can be hurt easily." In the words of one adolescent girl: "My mother can tell when I'm unhappy and 'play like' I'm happy." Others replied: "Exactly what I'm feeling inside. Sometimes she (mother) can see right through me." "They may know I'm worried or wondering about something when others don't." "Well, she knows how I feel about things—things I would not be able to discuss with anyone else."

As Table 8–1 shows, older children differ strikingly from younger ones in this regard. Whereas 13 percent of the 8–9 year olds answered in terms of such general feelings or thoughts, this was true of fully 41 percent of those 16 years or older—a ratio of over 3:1.

A second aspect of the psychological interior deals with *interper-*

*sonal thoughts and feelings.* Sometimes reference is made to feelings about specific others, at other times to feelings about general categories of others. Such information is often carefully concealed from all but the closest confidante. With reference to specific others, one high school girl attributed to the person who knew her best knowledge of "the way I feel about other people. Like my mother wouldn't know the way I really feel about her." More often the responses refer to categories of others: "my feelings about friends or enemies," "certain people I don't like," "my real personal feelings about people." Such interpersonal attitudes are clearly age-related; overall 12 percent of the older subjects but 3 percent of the younger cited inner thoughts and feelings about other people as their first answer.

Finally, older children were slightly more likely to mention certain *private or intimate aspects of the self,* such as personal desires, secrets, or aspirations. Although the trend is uneven, in general such responses are slightly more characteristic of those 14 or older than the rest. In a number of cases, these thoughts were so private that the subject refused to reveal them to the investigator—"I'd rather not say." For various reasons (embarrassment, danger of conflict, or other possible repercussions) the individual is cautious about revealing such thoughts and feelings to others.

What is reflected in the above responses, then, is a conception of the self defined in terms of an *inner world of thought and experience,* a world to which access is strictly limited. This world consists of a melange of inner thoughts, feelings, and wishes (usually expressed in general terms); interpersonal feelings and attitudes; and certain personal desires, aspirations, or secrets. Combining these three response categories, we find that 57 percent of those 16 years or older, but only 21 percent of those 8–9 years of age, utilized one of these three categories as their first response. (Discrepancies between cells and subtotals in Table 8–1 are due to rounding.) In sum, when asked what the person who knows him best knows about him that others do not, the older child, like the rest of us, is strikingly more likely than the younger to answer in terms of this psychological interior.

If the adolescent's responses tend to focus on inner thoughts, feelings or emotions, on what aspect of the self does the younger child focus? Overwhelmingly, the data show, he tends to respond in terms of *overt, especially behavioral, characteristics*—in other words, in terms of a social exterior.

*TABLE 8-1.

**What the Person Who Knows You Best Knows That Other People Do Not,
by Age (First Characteristic Mentioned)**

| | | | Age | | |
|---|---|---|---|---|---|
| | 8-9 | 10-11 | 12-13 | 14-15 | 16+ |
| *What person who knows you best knows about you* | | | | | |
| 1. Inner thoughts and feelings (general) | 13% | 21% | 28% | 38% | 41% |
| 2. Interpersonal attitudes | 3 | 5 | 10 | 10 | 12 |
| 3. Dreams, problems, secrets | 5 | 2 | 3 | 7 | 5 |
| *Psychological interior* (total %)† | 21% | 27% | 41% | 55% | 57% |
| 4. Behavior, conduct | 34 | 40 | 26 | 20 | 19 |
| 5. Abilities, achievements | 12 | 4 | 2 | 1 | 1 |
| 6. Physical characteristics | 7 | 4 | 5 | 3 | 1 |
| 7. Objective, demographic, material | 5 | 2 | 1 | 1 | 2 |
| 8. Interests, activities (specific) | 15 | 8 | 6 | 3 | 5 |
| *Social exterior* (total %) | 73% | 58% | 41% | 29% | 28% |
| 9. Other traits or dispositions | 3 | 7 | 7 | 3 | 5 |
| 10. Interests, likes (general) | 3 | 7 | 11 | 13 | 11 |
| *Other* (total %) | 5% | 14% | 18% | 16% | 16% |
| N = 100% | (149) | (244) | (254) | (208) | (240) |

†Discrepancies between cells, sub-totals and grand totals due to rounding.

*See footnote 7, chapter 3, p. 98.

Specifically, the single most common response to this question among the younger children deals with some kind of *behavior*. Unlike the older child, who believes that the person who knows him best has succeeded in penetrating to his psychological interior, the younger child believes that the person who knows him best is the one with restricted information about his *actions*. These actions tend to have a highly moral cast. Disapproved or punishable behavior is mentioned with particular frequency. Thus, when asked what the person who knows her best knows that other people do not, one little girl answers: "She (my mother) knows I cross (Bradley) Road when I shouldn't but she doesn't tell my father because he gets so mad. I tell her but not my father. She's the only one who knows I let some of my friends ride my bike when I'm not really supposed to." Another little boy replies: "That I watch adult shows more than kid shows." "He (father) knows how many crackups I've had on my bike. He knows why I come home from school late—which I do." Others replied: "that I'm bad, get into trouble," "roughneck," "what I do at parties that my mother doesn't like me to do," "when I take off my school clothes, I don't hang them up," "don't do no work in the house," "play hooky in school," "steal," "smoke," "I fight sometimes," "I tell stories," "me and my best friend get into trouble fighting."

To the younger child, then, the self is an actor who engages in certain kinds of behavior, much of which is morally evaluated—crossing streets, riding bikes, fighting, leaving messy rooms. Although some aspect of behavior—approved, disapproved, or neutral—is frequently selected by older children, the tendency is much stronger among the younger children, 35 percent of the 8–9 years olds answering in terms of behavior (a figure increasing till the age of 10–11), compared with 19 percent of those 16 years or older. In sum, the younger child is more likely than the older to describe the self in *behavioral, moral,* and *specific* terms.

Second, in response to this question younger children are also more likely to focus on *abilities or achievements*. What the person who knows him best knows about him is "how smart I am," "smartest person in the family," "good in arithmetic and bad in reading," "she knows about my reading and division," "my report cards are very good," "knows whether I'm going to pass or fail." Asked what the person who knows her best knows that others do not, one little girl replies: "She knows I can skate. I can read well." An elementary school boy: "She knew that I was a better baseball player." The primary answer of 12 percent of the youngest but 1 percent of the oldest was in terms of some skill, ability, or achievement (or its lack). These aspects of the self are, again, overt and visible.

Third, younger children are more likely than older to identify themselves in terms of physical characteristics (height, appearance, hair color, health status) or social identity elements (sex, age, race, class). These characteristics are relatively objective and factual and most are visible aspects of the self. Sometimes the child says that the special knowledge about himself vouchsafed to the other is some aspect of his *appearance*: "what I look like," "what I wear," "I'm tall," "short," "fat," "my clothes." At times the knowledgeable other —usually the mother—is a health expert. This person knows "if I'm sick," "when I don't feel good," "when I'm sick, she knows what's wrong even though I can't explain it." The importance of visibility is apparent in the response of the little boy who replied: "They know I've got stitches in my head, and no one knows it except my family." Finally, these children sometimes answer in terms of an assortment of material and demographic facts about the self: "how much money I have," "my real age," "my background," "that I have a new pair of shoes." In all, 12 percent of the youngest but only 3 percent of the oldest mention one of these physical characteristics or social identity elements.

Finally, the younger children show a greater tendency to mention their preferences for, or interest in, certain objects or activities. Since preferences or interests refer to inner states of thought and feeling, this finding may appear to contradict the claim that the young child tends to think of the self in terms of a social exterior rather than a psychological interior. As indicated in the discussion of self-attribution theory, however, many preferences are most conspicuously evidenced, and most clearly recognized, by overt behavior. Thus, even in expressing something so apparently subjective as likes and dislikes, the young child remains to a considerable extent an objectivist. In the words of an elementary school boy: "He (the little boy up the street) knows what games I like. He knows what kinds of work I like to do, what my favorite subject is." Or consider the preferences of the little girl who tells us what her father knows about her that others do not: "My father knows I like to buy candy. My mother doesn't know this." Other children reply: "She knows what kinds of things I like, what I like to do, to work with, to fool around with, to make." "She knows I like music." "She knows places I like to go to best." "She knows I like hatchets and cuttin' tools." "Well, my father knows the sports I like, and he knows I like to build things." "That I don't like to eat very much." "Like to read or dance." "Like to go to ball games." Fifteen percent of the 8–9 year olds but 5 percent of those 16 and over responded in these terms.

Thus, when the young child is asked what the person who knows him best knows that others do not, he tends more than the older to answer in terms of certain overt, visible, external, and public categories. His first reply will either be in terms of his behavior (especially good or bad behavior); his talents, abilities, or accomplishments; certain objective (physical or social identity) characteristics; and certain likes or dislikes, most of which have overt manifestations. When we combine these four general categories of response, we find that *73 percent of those 8–9 years of age, but only 28 percent of those 16 or over, responded in terms of one of these external characteristics.*

To summarize, when asked what the person who knows him best knows that others do not, the older child tends to answer in terms of a psychological interior—a world of general emotions, attitudes, wishes, and secrets—while the younger child is more likely to respond in terms of a social exterior—a world of behavior, objective facts, overt achievements, and manifested preferences. The younger child's view is turned outward, toward the overt and visible; the older child's gaze is turned inward, toward the private and invisible.

Self-concept development, then, would appear to follow an extremely interesting course. As the child grows older, he becomes less of a demographer, less of a behaviorist, more of a psychological clinician. Expressed in broadest terms, *with increasing age the child becomes less of a Skinnerian, more of a Freudian.*

Although the question of what "the person who knows you best knows that others do not" reveals certain things about how the child spontaneously conceptualizes the self, it also obscures others. In response to this question, a child is unlikely to say "boy," "black," "American," or "big," not because he does not conceptualize the self in these terms, but because even someone who knows him slightly knows these facts about him as well as the person who knows him best of all. For example, our mother knows we like grilled cheese, but our teacher doesn't know this; our teacher knows we are poor at spelling, but our friend does not know this. Friend, teacher, and mother, however, all know whether we are a boy or girl, black or white, tall or short, blonde or brunette, old or young. When we turn to the various other questions about the self cited above, therefore, the substantive answers naturally differ considerably. But, as we shall see, they are entirely consistent with the above observations, reinforcing the major conclusions and adding others.

PRIDE, DISTINCTIVENESS, DESIRED SELF

Tables 8–2 to 8–6 examine the relationship between age and *points of pride* ("what things are really best about you, are the best part of you") (Table 8–2); *points of shame* ("any weak points, that is, any things not as good about you") (Table 8–3); *sources of distinctiveness* ("in what ways are you different?") (Table 8–4); *sources of commonality* ("in what ways are you the same?") (Table 8–5); and *desired adult self* ("what kind of personality would you like to have when you grow up? I mean, what kind of person would you like to be?") (Table 8–6). In order to maximize comparability, the answers are classified according to the following response categories wherever possible: (1) physical characteristics (2) social identity elements (3) behavior (4) abilities or activities (5) interpersonal traits, and (6) other traits.

Instead of examining each question separately, we have elected to look at the general pattern of responses in terms of the conceptualization of the self as either a social exterior, or a psychological interior. These two modes of self-conceptualization will be considered separately.

THE SELF AS SOCIAL EXTERIOR

What is it about himself that the child is most proud of—"really best about you, . . . best part of you?" Of all the thousands of things about the self that younger children might choose, it is striking how many of them describe the best thing about themselves as some *physical or bodily characteristic*. To many elementary school children, the best part of the self is "my skin," "my hands and my teeth," "my eyes," "my head, my feet, and my arms," "I like the color of my eyes," "my face, my hands, and the way I'm shaped," "my hair, my clothes, my shoes," "my wrist. I show people I am strong, and my muscle." Younger children are also somewhat more likely to mention social identity elements (boy-girl, old-young, black-white), though few children center their sense of pride in such demographic properties. As Table 8–2 shows, 32 percent of the youngest respondents, but only 8 percent of the oldest, cited a physical characteristic as the "best thing" about him.

**\*TABLE 8-2.**

*Chief Points of Pride, by Age*

| | Age | | | | |
|---|---|---|---|---|---|
| | 8-9 | 10-11 | 12-13 | 14-15 | 16+ |
| *Things really best about you* | | | | | |
| Physical Characteristics | 32% | 28% | 18% | 13% | 8% |
| Social Identity | 1 | 1 | 3 | 1 | — |
| Behavior | — | — | — | — | — |
| Abilities-Activities | 47 | 44 | 45 | 38 | 28 |
| Interpersonal Traits | 9 | 14 | 11 | 17 | 28 |
| Other Traits | 12 | 14 | 23 | 32 | 36 |
| N = 100% | (206) | (330) | (293) | (245) | (286) |

*See footnote 7, chapter 3, p. 98.

**\*TABLE 8-3.**

*Chief Points of Shame, by Age*

| | Age | | | | |
|---|---|---|---|---|---|
| | 8-9 | 10-11 | 12-13 | 14-15 | 16+ |
| *Weak points* | | | | | |
| Physical Characteristics | 22% | 27% | 18% | 15% | 10% |
| Social Identity | — | — | — | — | — |
| Behavior | — | — | — | — | — |
| Abilities-Activities | 58 | 52 | 55 | 44 | 41 |
| Interpersonal Traits | 6 | 6 | 6 | 8 | 17 |
| Other Traits (chiefly self-control) | 14 | 15 | 20 | 33 | 32 |
| N = 100% | (106) | (201) | (236) | (201) | (189) |

*See footnote 7, chapter 3, p. 98.

What about himself is the child *least* proud of; that is, what does he consider the weakest part about the self? Again, 22 percent of the youngest but 10 percent of the oldest cited some physical character-istic: "My legs," "no nice shape," "not strong enough." Virtually none of the respondents cited a social identity element (Table 8–3).

Asking the child to indicate in what ways he is *different* from others and in what ways he is the *same* affords the child another opportunity to conceptualize the self spontaneously. Asked in what ways he is "different from other kids you know?" the young child again sees his chief source of distinctiveness as resting in the most overt and visible aspects of the self—chiefly physical aspects. "Some children around me have short hair and I don't," "I'm bigger than they are," "some are skinny," "I have different color eyes, hair, and skin," "other people might have sores or marks on them," "I dress different," "they don't look like me," "some boys are lighter than me." Also mentioned by the younger children, though to a considerably lesser extent, are certain of the most visible elements of social identity. "I am a boy." "They might be older than me." "Some I know are Chineses." "By my looks and name. One other person got my name—my cousin." As Table 8–4 shows, older and younger children differ strikingly in their ideas about what distinguishes them from others. Over four times as many of the youngest as of the oldest (38 percent to 8 percent), asked to described in what ways they were different from others they knew, answered in terms of some physical or material characteristic, or some closely allied aspect of social identity such as age or sex.

Similar results appear when subjects are asked: "In what ways are you the same as most other kids you know? (Are there any other ways?)." The points of similarity that the young child sees between himself and others again turn out to be conspicuously physical characteristics and social identity elements. The younger child is more likely to say that "we look the same," "we dress the same," "have the same kind of clothes," "we're all small," "we're all alive," "we all breathe," "we all sleep," or else that "we're all the same age," "we go to the same church," "we are the same race," "I am a boy and so are my friends," "we're all girls." Thirty-seven percent of the youngest children, but only 17 percent of the oldest, selected as their point of similarity with others either some physical characteristic or some social identity element (Table 8–5).

The final open-ended question, dealing with the desired adult self, is somewhat ambiguous, but the results are worth reporting. "Now, I would like to ask you what kind of person you would like to be when you are an adult. What kind of life would you like to lead? How would you like to spend your time? What kinds of things would you like to do when you are an adult?" This question was followed by other questions dealing with "What kind of home would you like to have when you're an adult?" "How would you like to look when you're an adult?" and finally, "What kind of personality would you like to have when you grow up? I mean, what kind of person would you like to be?"

One problem with the last question, of course, is that the younger children may not understand the term "personality." We tried to

**\*TABLE 8-4.**

*Sources of Distinctiveness, by Age*

|  | Age | | | | |
|---|---|---|---|---|---|
|  | 8-9 | 10-11 | 12-13 | 14-15 | 16+ |
| *How different from others?* | | | | | |
| Physical Characteristics | 33% | 30% | 19% | 10% | 7% |
| Social Identity | 5 | 3 | 4 | 4 | 1 |
| Behavior | 15 | 22 | 17 | 19 | 19 |
| Abilities-Activities | 29 | 27 | 37 | 33 | 26 |
| Traits | 17 | 17 | 22 | 34 | 46 |
| N = 100% | (144) | (241) | (279) | (211) | (259) |

\*See footnote 7, chapter 3, p. 98.

**\*TABLE 8-5.**

*Sources of Commonality, by Age*

|  | Age | | | | |
|---|---|---|---|---|---|
|  | 8-9 | 10-11 | 12-13 | 14-15 | 16+ |
| *How same as others?* | | | | | |
| Physical Characteristics | 26% | 13% | 13% | 12% | 10% |
| Social Identity | 11 | 8 | 8 | 8 | 7 |
| Behavior | 5 | 9 | 7 | 6 | 5 |
| Abilities-Activities | 55 | 60 | 65 | 64 | 68 |
| Traits | 4 | 11 | 6 | 9 | 10 |
| N = 100% | (151) | (288) | (290) | (249) | (247) |

\*See footnote 7, chapter 3, p. 98.

deal with this problem by asking about "what kind of person" the child would like to be, but this expression may not have the identical meaning.

Since the immediately prior question had asked the respondent to indicate how he would like to *look* as an adult, this fact effectively precluded answers to these questions in terms of physical characteristics. Keeping this context in mind, it may be noted that the younger children were much more likely to answer the question in terms of social identity elements—groups, roles, statuses, or social categories. They would like to be a good mother, a family man, nice to my children, a fireman, a baseball player, a good Catholic. Occupations are mentioned frequently: "I would like to be a teacher," "A salesman or something," "Doctor," "A policeman." Sometimes fantasy occupations are mentioned: "I would want to act like a cowboy and blow the Indians' heads off," "A man who plays baseball, a good player." The family role most often mentioned is "mother"— "a mother," "a lady." Fully 33 percent of the youngest, but only 2 percent of the oldest, offered a response in terms of such social identity elements (Table 8–6).

**\*TABLE 8-6.**
*Desired Adult Personality, by Age*

|  | Age | | | | |
|---|---|---|---|---|---|
|  | 8-9 | 10-11 | 12-13 | 14-15 | 16+ |
| *Person like to be when grown* | | | | | |
| Physical Characteristics | — | — | — | — | — |
| Social Identity | 33% | 20% | 8% | 4% | 2% |
| Behavior | 2 | 4 | 4 | 4 | 2 |
| Abilities | 8 | 12 | 11 | 4 | 3 |
| Interpersonal Traits | 36 | 46 | 58 | 69 | 73 |
| Other Traits | 21 | 18 | 19 | 19 | 20 |
| N = 100% | (107) | (192) | (201) | (180) | (208) |

*See footnote 7, chapter 3, p. 98.

In sum, although the specific content of the answers to these unstructured questions naturally differ, the propensity of younger children to conceptualize the self in terms of a social exterior is striking. Asked about the best thing about them; what they are most proud of; what things about them are not so good; in what ways they differ from others; in what ways they are the same as others; and what kind of person they would like to be as an adult, the younger children are much more likely to focus on the most overt and visible aspects of the self—physical characteristics and, to a lesser extent, social identity elements.

In order to highlight the extraordinary propensity of younger children, when exposed to open-ended questions, spontaneously to conceptualize the self in terms of the elements of a social exterior, Table 8–7 indicates whether respondents of various ages answered first in terms of either a physical characteristic or social identity element or whether they initially selected some other response. It is immediately apparent that strong and consistent changes in modes of self-conceptualization take place as children grow older.

**TABLE 8-7.**

*Self as a Social Exterior, by Age*

| | Age | | | | |
|---|---|---|---|---|---|
| | 8-9 | 10-11 | 12-13 | 14-15 | 16+ |
| *Best thing about you* | | | | | |
| Physical characteristics | 32% | 27% | 18% | 13% | 8% |
| Social identity | – | 1 | 3 | 1 | – |
| Other | 68 | 72 | 79 | 86 | 92 |
| N = 100% | (206) | (330) | (296) | (245) | (286) |
| *Things not as good* | | | | | |
| Physical characteristics | 22% | 27% | 18% | 15% | 10% |
| Social identity | – | – | – | – | – |
| Other | 78 | 73 | 82 | 85 | 90 |
| N = 100% | (106) | (201) | (236) | (201) | (189) |
| *Different from others* | | | | | |
| Physical characteristics | 33% | 30% | 19% | 10% | 7% |
| Social identity | 5 | 3 | 4 | 4 | 2 |
| Other | 62 | 67 | 77 | 86 | 91 |
| N = 100% | (144) | (241) | (279) | (211) | (259) |
| *Same as others* | | | | | |
| Physical characteristics | 26% | 13% | 13% | 12% | 10% |
| Social identity | 11 | 8 | 8 | 8 | 7 |
| Other | 63 | 79 | 79 | 80 | 83 |
| N = 100% | (151) | (288) | (290) | (249) | (247) |
| *Personality as adult* | | | | | |
| Physical characteristics | – | – | – | – | – |
| Social identity | 33% | 20% | 8% | 4% | 2% |
| Other | 67 | 80 | 92 | 96 | 98 |
| N = 100% | (107) | (192) | (201) | (180) | (208) |

*See footnote 7, chapter 3, p. 98.

# From Percept to Concept

TRAITS

The propensity to think of the self in terms of a trait psychology is not there at birth but is, rather, the product of social development.

The self perceived by the young child is largely a concrete, material reality. With maturation and learning, however, the individual comes to conceptualize the self in terms of more abstract *response tendencies or potentials*, consisting largely of dispositions or traits. In speaking of the "development of a terminology of trait names and similar abstractions," Gardner Murphy (1947: 505–6) notes that in time "the vocabulary of the self becomes, so to speak, less and less visual and in general less and less sensory. . . . The child forms general ideas of himself. In short, the self becomes less and less a pure perceptual object, and more and more a conceptual trait system."

Table 8–8 summarizes the proportions of subjects at various age levels who answered the various questions in terms of traits. With regard to three of the questions (best thing about you, things not so good, kind of person as adult), it proved possible to differentiate *interpersonal* traits from other types; but with regard to sources of distinctiveness or commonality (different from others, same as others), it was not possible to make this distinction, usually because an insufficient number of people mentioned interpersonal traits. An examination of each question will provide a more specific idea of older and younger children's modes of self-conceptualization.

Consider the responses to the question: "What kind of personality would you like to have, or what kind of person would you like to be, when you are an adult?" Whereas 58 percent of the youngest children cited some trait, this was true of 93 percent of the oldest. These traits were varied in nature. Some referred to desired emotional characteristics (cheerful, happy), to matters of emotional control (doesn't lose temper, avoids fights), to qualities of character (mature, brave, dependable, honest), to interpersonal characteristics (friendly, outgoing, sociable, well-liked, kind, considerate), and others. The answers, it should be stressed, were coded in terms of meanings, not words. Typical responses were: "I'd like to have people like me, get along good with other people." "Fun and friendly." "A happy-go-lucky personality." "Kind that would get along with most." "I wouldn't want to worry much." "Serious, but know when to have fun." "Pleasant and sincere." "I'd like to be good-natured. Maybe have a little bit of a mean streak." "The kind that doesn't give trouble to anybody." "A person who understands other people, helps other people."

Concerning points of pride ("the best thing about you"), the older children are again strikingly more likely than the younger to

**TABLE 8-8.**

*Self-Concept as a System of Traits, by Age*

| | Age | | | | |
|---|---|---|---|---|---|
| | 8-9 | 10-11 | 12-13 | 14-15 | 16+ |
| *Best thing about you* | | | | | |
| Interpersonal traits | 9% | 14% | 12% | 17% | 28% |
| Other traits | 12 | 14 | 23 | 32 | 36 |
| Other responses | 79 | 72 | 65 | 51 | 36 |
| N = 100% | (206) | (330) | (296) | (245) | (286) |
| *Things not as good* | | | | | |
| Interpersonal traits | 6% | 6% | 6% | 7% | 18% |
| Other traits (self-control) | 14 | 15 | 20 | 33 | 32 |
| Other responses | 80 | 79 | 73 | 60 | 50 |
| N = 100% | (106) | (201) | (236) | (201) | (189) |
| *Different from others* | | | | | |
| Traits | 18% | 17% | 22% | 34% | 46% |
| Other responses | 82 | 83 | 78 | 66 | 54 |
| N = 100% | (144) | (241) | (279) | (211) | (259) |
| *Same as others* | | | | | |
| Traits | 4% | 11% | 6% | 9% | 10% |
| Other responses | 96 | 89 | 94 | 91 | 90 |
| N = 100% | (151) | (288) | (290) | (249) | (247) |
| *Personality as adult* | | | | | |
| Interpersonal traits | 36% | 46% | 58% | 69% | 73% |
| Other traits | 22 | 18 | 19 | 19 | 20 |
| Other responses | 42 | 36 | 22 | 12 | 7 |
| N = 100% | (107) | (192) | (201) | (180) | (208) |

*See footnote 7, chapter 3, p. 98.

consider the best thing about them to be some trait. Fully 64 percent of the oldest, but 21 percent of the youngest, cited a trait as the best thing about them. "The way I apply myself." "I stay out of trouble." "I don't lose my temper easily." "The ability to think, analyze a situation, and come up with a good answer." "Sometimes I can take an unserious attitude when something's too serious. I don't think I laugh at it, but just not feel so badly." "I supposedly have common sense." "My ability to try hard enough." "Have a good philosophy." "I don't get mad easily. I'm not mean to people who are different." "Trying to make the best out of a situation." "My attitude—if I don't like something, I won't get nasty about it."

When asked about their drawbacks ("Weak points . . . things that aren't as good about yourself"), the older subjects' propensity to conceptualize the self in terms of traits again appears. Fully 50 percent of the oldest, but only 20 percent of the youngest, located their chief deficiencies in this area.

Similar results appear when we turn to the respondent's view of his sources of distinctiveness—in what ways he feels he is "different from most other kids you know." Forty-six percent of the oldest subjects, but only 18 percent of the youngest, cited some trait as that characteristic which particularly set them off from others they knew (Table 8–8). But asked in what way they were *the same* as other kids they knew, very few subjects of any age answered in terms of some trait or personality quality (only 10 percent of the oldest and 4 percent of the youngest); nor is any particular age trend discernible. In general, youngsters do not see themselves as similar to others in terms of underlying dispositions but in terms of behavioral norms— going to parties, listening to rock, enjoying riding, listening to records—in other words, conforming to peer group practices. In the adolescent's view, he *differs* from his peers in basic personality qualities but is *like* them in his overt behavior or interests.

### INTERPERSONAL TRAITS

Although the self-concepts of older children are far more likely to be composed of traits, the most striking difference lies in the adolescent's tendency to conceptualize the self as a language of *interpersonal* traits. Although it is hard to draw a sharp line between interpersonal and other traits (chiefly because our other traits may affect how we get along with others), for present purposes we shall focus on three types of interpersonal traits. The first type includes those traits describing the individual as *attracted* to others. He thinks of himself as "friendly," as someone who "likes to be with people," who "enjoys people," "likes others." They are defined by the individual's general feelings *toward* other people. The second category involves descriptions of the individual as *attractive* to others, as someone toward whom positive feelings are felt. These would include such terms as "well-liked," "easy to get along with," "pleasant," "popular," "well-respected," "winning," or "pleasing" personality. Here the individual defines himself in terms of others' characteristic reactions to him; one might describe these as "looking glass" traits. The third category refers to what might be described as "interpersonal virtues." These traits consist largely of prescriptions for desirable interpersonal behavior—one *should* be cooperative, kind, helpful, sharing, unselfish. Although such desirable behavior implies corresponding positive feelings, such feelings are not indispensable.

Does the tendency to conceptualize the self as a language of interpersonal traits change as children grow older? With regard to the three questions in which it proved possible to distinguish interpersonal traits from other traits, Table 8–8 indicates that the older subjects are considerably more likely than the younger to conceptualize the self in these terms. Asked what kind of personality they would like to have, or what kind of person they would like to be, as an adult, fully 73 percent of the oldest but only 36 percent of the youngest cited some interpersonal trait. Similarly, 28 percent of the oldest but only 9 percent of the youngest cited an interpersonal trait as their chief point of pride ("best thing about you"). When asked about their chief deficiency or drawback, 18 percent of the oldest cited some interpersonal trait or some interpersonal difficulty (trouble with someone), whereas only 6 percent of the youngest answered in these terms. With regard to similarity or difference from others, too few children of any age mentioned interpersonal traits to warrant comparisons.

The second point is that the prime difference between older and younger subjects appears to be with regard to *interpersonal attraction traits*—liking and being liked by others—whereas *interpersonal virtues* are less clearly related to age. The most striking difference is apparent with reference to desired adult characteristics: 59 percent of the oldest but 6 percent of the youngest want chiefly to like others or to be liked by them. Thus, when asked what kind of personality they would like to have or what kind of person they would like to be as an adult, high school students replied: "a person that's well-liked by most people," "well-liked—try to think of others," "one that gets along with about everybody," "easy to talk to," "a nice person with a good reputation," "charming, outgoing, friendly, honest," "nice, sweet personality," "easy to get along with," "friendly person who can get along with just about everybody," "very friendly and active—sociable," "well-respected person," "I'd like to be admired."

Regarding chief points of pride—"the best thing about you"—23 percent of the oldest but 3 percent of the youngest cited an interpersonal attraction trait. Almost all of these turn out to reflect the degree to which the individual is attractive to others, not the degree to which he is attracted to them. This finding is understandable. The child may be proud that he is well-liked or gets along well with people —this is a tribute to his interpersonal skills or success—but it requires no special talent to like people or to have friendly feelings toward them. Thus, the older subjects tell us that the best thing about them is being "a good friend," "easy to get along with," that they have "ability to get along with different kinds of people," "I try to put people at ease," "My sense of humor, making friends." "It doesn't take me long to get along with people." "My personality." "I get along well with people—that's an important thing in life." "My personality, people say. They tell me the best part about me is when I smile." "Yes—gee whiz! I can't think of anything specific. Friendly— please my family, be nice to them." "I have an ability to make friends, a stronger ability than most people."

IMPULSE OR EMOTIONAL CONTROL

No task of socialization is more fundamental than the teaching of self-control. Every society must somehow arrange to regulate the expression of spontaneous impulses, particularly those of sex and aggression. Hence, one of the first tasks of socialization is to prevent the child from doing anything he feels like doing—painting the walls, running in the street, clubbing his baby brother, eating all the cookies in the package, playing with matches. Indeed, the need for such control continues throughout life, though the wants change. Such controls are imposed both externally, through punishment, and internally, through conscience. Self-control, then, is the victory of superego over id, of internalized rules of behavior over biopsychological urges.

Although young children probably have weaker self-control than older ones, the data suggest that self-control is experienced as a greater *problem* by older children. In speaking of their chief drawbacks or deficiencies—"things not so good about you"—nearly a third of the older children mentioned some self-control problem compared to a sixth of the younger subjects. The chief weaknesses cited by high school students are: "Sometimes I get mad at my sister over nothing." "It's hard for me to force myself to do homework or whatever I'm supposed to do." "I get in fights with my parents." "Too frank most of the time." "When someone insults my feelings I cry a lot." "Getting upset too easily." "Short temper."

Impulse control, of course, is a major problem with younger children as well,[1] but it is apparently more likely to be *conceived* as a self-deficiency by the older (Table 8–8). One conclusion to be drawn is that older children are more *aware* of an inner life of wish, desire, and impulse and of themselves as agents struggling to control and restrain it. The adolescent is better able to objectify the self, to stand outside the self and to attempt to tame these impulses; and, failing this, to feel chagrin, regret, or shame at the self.

It is worth noting that, although defeat by urges and impulses is ego-diminishing, victory over them is not correspondingly ego-enhancing. In other words, whereas nearly a fourth of the subjects who gave meaningful answers cited some failure to control impulses as their chief weakness, almost none cited success in impulse control as their main strength. The individual reproduces internally the treatment meted out by society; infraction of rules is specifically punished, but conformity to them is merely taken for granted.

## Discussion

The aim of this analysis has been to learn how younger and older subjects spontaneously organize and structure the self-concept when exposed to diverse general questions about the self. Varied aspects of the self-concept emerge. There is, first, a physical and material self, apprehended by the eye, visible and palpable. It consists of the individual's body, features, appearance, attractiveness, and size. Second, it is a social structure—the elements of social identity—which represents the categories to which the individual is socially recognized as belonging. These include age, sex, race, religion, nationality, etc. Third, there is the self as social actor, defined by what it characteristically does—plays baseball, goes to dances, listens to music, smokes pot. Fourth, it is a self characterized by certain abilities or talents; levels of competence, achievement, or skill characterize all realms of human activity and weigh heavily in the individual's assessment of personal worth. Fifth are certain interests and attitudes. The individual sees himself as someone who is interested in philosophy, likes classical music, is concerned with health foods, enjoys movies, follows sports. Sixth, there is a set of abstract characteristics which can loosely be subsumed under the general heading of personality traits. These exist in endless variety and abundance, defying any simple classification. One basic distinction forces itself to attention, however—that between interpersonal traits and other traits. Finally, closely allied with traits—and at times indistinguishable from them —are inner thoughts, feelings, and attitudes. These are not only part of the self, but of the self-concept as well.[2]

These patterns of self-conceptualization are not essentially different from the ways people conceptualize others. As Shantz (1975: 314) observes: "There is a developmental trend toward conceiving of people less in terms of their surface appearance, possessions, and motor behavior and more in terms of an underlying reality"; for example, values, beliefs, and intentions (See also Flapan, 1968). Certain cognitive processes govern how all people, including the self, are conceived.

In stressing that older children are much more likely to conceptualize the self in terms of inner thoughts and feelings or within the framework of interpersonal and other traits, we do not mean to suggest that they do not *also* see themselves as material or physical entities, that they are not keenly and acutely aware of their elements of social identity, that they do not perceive themselves as actors, or that they are not disposed to characterize themselves in terms of abilities, interests, or activities. They are, in fact, disposed to do all these things. If the older child spontaneously cites a trait or inner feeling, then, it is not because the other components are not also important elements of his self-concept, but either because they are taken for granted, have lower priority or centrality in his scale of values, or are felt to be less of what he truly and deeply feels himself to be.

In reflecting on these findings, two questions are especially intriguing: (1) why are adolescents so much more likely than younger children to conceptualize the self in terms of a psychological interior (an inner world of thought and feeling); and (2) why are adolescents so much more likely to conceptualize the self in terms of traits, particularly interpersonal traits? These questions will be considered separately. The basic explanation of the age differences, we believe, is to be found in the different cognitive processes of children and adolescents. We suggest that three such processes critically differentiate children and adolescents and importantly contribute to self-concept differences: introspection, communication, and conceptualization.

## INTROSPECTION

On initial reflection, the finding that the young child is apparently so oblivious to a psychological interior—an inner world of thought and feeling—may appear puzzling. It is the younger child, after all, who is fundamentally egocentric (Piaget, 1948), who lives in his own little world, who sees matters entirely from his private perspective, and who is less able to adopt a detached and objective attitude toward the world. The older child, on the other hand, is less wrapped up in his own private world, more attuned to external reality. We are thus faced with the anomalous fact that the young child, living in his own world and seeing matters largely from his own special point of view, is essentially unconscious of an inner world of thought and experience. The older child, able to adopt a more objective, detached view, tends more to think of himself in terms of an unobservable, or not readily observable, self. In order to understand this paradox, it is necessary to have recourse to an important but neglected concept in psychology, namely, the concept of introspection.

Webster tells us that to introspect is "to look into or within, as one's own mind; to inspect, as one's own thoughts and feelings; to practice self-examination." What is actually involved in looking into one's own mind or inspecting one's own thoughts and feelings? For example, assume we ask someone who has solved a problem how he reached this solution. What we are actually asking him to do is to reflect on his own thought processes—to tell us what is going on in his mind or, in retrospection, what went on in his mind.

With respect to the issue of children's self-concepts, we suggest that the decisive reason why the older child tends to view himself in terms of a psychological interior while the younger does not is that the former is more capable of introspection. The older child is disposed to conceptualize himself as a being who thinks and feels. The younger child, on the other hand, is relatively unaware of a psychological interior because he has not developed the tendency to introspect, i.e., to look into his own mind or inspect his own feelings.

Why should this be? In order for introspection to occur, one essential precondition must be satisfied: *thought must first become conscious of itself*. This process, so characteristic of the mature mind, is rare among younger children. In Piaget's (1928: 12, 137) words:

"There is nothing in egocentrism which tends to make thought conscious of itself. . . ." The young child lacks "the habit of watching himself think, i.e., of constantly detecting the motives that are guiding him in the direction he is pursuing."

The introspection process, such a constant feature of adult consciousness, is actually a remarkable and mysterious phenomenon. As Murphy (1947) observed, Descartes, seeking logical evidence of his existence, rooted it in his thought processes: "I think, therefore I am." What he failed to appreciate was the remarkable fact that he could stand outside himself and think of himself as a thinker.

Unlike the young child, the mature individual is alert and responsive to his own thought processes. Examples come easily to mind. We think of an idea and dismiss it as stupid; we have feelings of greed or lust and feel ashamed of these urges; we wish we felt more enthusiastic about our jobs; we laugh heartily at our own jokes, argue with our own conclusions, deplore our own excitability, are chagrined at our own stupidity. In other words, we have thoughts and feelings about our thoughts and feelings. The reason that the older child is more aware of a psychological interior is that he is capable of introspection, of reflecting on, and reacting to, his inner cognitive, affective, and conative processes.

These data, showing that the younger child tends to conceive of the self as a social exterior, the older as a psychological interior, would imply that the older child is somehow closer to his real self, the younger child more alienated from it. We believe the contrary is true. We are not suggesting that the young child does not *have* an inner life of thought and feeling but rather that he does not *reflect* on an inner life of thought and feeling. In other words, there is a fascinating but critical distinction between thinking—the spontaneous processes occurring in the mind of the individual—and thinking about thought. The young child, like the rest of us, experiences spontaneous thought —that is, certain images, words, fragments of ideas, perceptions cross his mind constantly. What the young child does less than others, however, is to think about thought, to reflect on his psychic or intellectual processes, to have feelings about his feelings, wishes about his wishes.

In sum, the reason the younger child is less likely to see himself in terms of a psychological interior, we believe, is that he has not fully developed the tendency to introspect. The young child is a radical empiricist, an objectivist, a behaviorist; he responds to matters and actions which strike the eye. The older child is more of a psychological clinician; equipped with the ability to introspect, he is able to reflect on an inner world of thought, feeling, and wish. This difference in modes of self-conceptualization between older and younger children is profound.

But this conclusion, while answering one question, raises another: why is it that younger children, wrapped up as they are in their inner lives, are nonetheless incapable of seeing into themselves? Is it simply a lack of intellectual maturation? Although maturation is implicated, the answer is actually more sociological: it is nothing more nor less than the fundamental social process of communication.

## COMMUNICATION

Indispensable to communication, or to mature social intercourse generally, is the ability to take the role of the other, to see matters from the viewpoint of others. Piaget has described the precommunicative stage of development as "egocentricity"—the tendency to see the world exclusively from one's own viewpoint. In Piaget's words: ". . . thought in the child is ego-centric, i.e., the child thinks for himself without troubling to make himself understood nor to place himself at the other person's point of view" (1928: 1). The reason the child tells us that "Benjy just had a bath" whereas the adult tells us that "Benjy, our dog, just had a bath" is that the child does not keep our point of view in mind when selecting his verbal symbols.

Piaget's (1928) example of the "brothers test" is an apt one. The little boy knows that he has two brothers but thinks that each of his brothers has but one. It is easy for him to see that Billy is his brother but hard to see that he is Billy's brother; he has difficulty in comprehending that the world, as Billy sees it, includes him. The little girl knows that Janie is fun to play with without realizing that, to Janie, she is fun to play with.

Sweet and Thornburg (1971) asked a sample of 3–5 year olds to create their own families by selecting three-dimensional abstract figures of varying sizes and placing them in a frame. Although the children could place and name the other family members, only 60 of the 120 included themselves in the family (despite the fact that the self had been made salient in previous tasks). The young child, it is evident, sees the family from his own point of view (they are his family), not from their view (he is part of their family).

Mead stressed that in order to communicate, it is essential to put oneself in the place of the other, to view matters from the perspective of the other. If we seriously wish to communicate, we cannot simply say what comes to our minds but must pay attention to *our own* words. If we wish to get a message across, we must put our thoughts in order, present them in a logical sequence, and anticipate others' responses to them. In order to do this, it is necessary to experiment with certain ideas, rejecting, selecting, and arranging them before presentation. Certain spontaneous thoughts remain unexpressed; others are modified in order to avoid eliciting an undesired reaction; still others are arranged so that they have the appearance of sense and plausibility. But if we must decide which of our thoughts to express and which to suppress, and if we must further reflect on alternative ways of arranging these thoughts so that they have the desired effect (keeping in mind the other's viewpoint), then one conclusion is inescapable: we must become responsive to *our own* thoughts.

It was one of George Herbert Mead's seminal insights to recognize that the individual, selecting his communications with a view to the others' response, at the same time reacts to his own words as if he were another. In Mead's (1934: 62–65) terms: "The importance of the vocal stimulus lies in the fact that the individual can hear what he says and in hearing what he says is tending to respond as the other person responds. . . . We can hear ourselves talking and the import of what we say is the same to ourselves that it is to others. . . . The vocal gesture, then, has an importance which no other gesture has. One hears himself when he is irritated using a tone that is of an irritable quality, and so catches himself." Elsewhere, Mead (141) states: "One starts to say something, we will presume an unpleasant something, but when he starts to say it he realizes it is cruel. The effect on himself of what he is saying checks him." Mead points out that we even respond to our own commands or wishes. We ask someone to get a chair, and the sound of our words arouses in ourselves the same inclination they arouse, or are intended to arouse, in others. This is no mere peculiarity or idiosyncrasy; it is, according to Mead, nothing more nor less than the nature of reflective intelligence, or, according to Piaget, the essence of human logic.

In sum, we suggest that, in order to communicate, we must first be able to take the role of the other, put ourselves in the others' shoes. But it is precisely the process of communication that gives birth in adolescence to introspection—the tendency to survey, assess and reflect on the inner world of thought, feeling, and desire. This is because we no longer simply think, but also think about our own thoughts and even think of ourselves as thinkers. The younger child, having no need to communicate, says anything that comes to his mind. Having less concern with whether he is understood or how the other will react, he has little need to pick and choose his words; hence, there is nothing to make him consciously aware of his own thought processes.

If this is so, then it gives rise to a generalization which, though seemingly paradoxical, is nonetheless true: *we must first get outside ourselves in order to be able to see inside ourselves.*[3] Both younger and older children are able to conceptualize the self in terms of a social exterior, but it is only the older children, attuned to the arts of communication and metamorphosed by the thought processes that spring from it, who appear able to conceptualize the self in terms of a psychological interior.

## CONCEPTUALIZATION

As noted above, the data strikingly support Murphy's *apercu* that, in the course of time, the self becomes less of a perceptual object and more of a conceptual trait system. Both maturation and learning are involved in this development. Why is the young child likely to think of the self as someone who rides a bike, likes chocolate ice cream, or plays Greek dodge ball and unlikely to characterize himself as kind, brave, lively, or cooperative? It is not a matter of vocabulary as such because the older children, in employing these concepts for self-description, rarely use these trait terms. Rather, they use descriptive phrases to express the concepts, many of which are within the younger child's range.[4] If the young child does not describe himself this way, we believe, it is not chiefly because he lacks the term, —although that may be a contributory factor—but because he has not mastered the concept.

What is a concept? Simmel (1950), we believe, has best captured the essence of the idea: "A concept isolates that which is common to diverse and heterogeneous items." In terms of his definition, elementary school children assuredly possess the power of conceptual thought, that is, the ability to isolate and distinguish that which is common to diverse and heterogeneous items. In time the child learns that a number of objects have certain features in common to which the term "chair" is applied, though these objects may differ in shape, color, size, composition, and so on.

Although the younger children in our sample have certainly mastered a large number of low-level concepts, what makes the higher-level concepts so much more difficult to grasp is that the underlying commonality of their manifestations is concealed by surface diversity. *Different* behavior may reflect the *same* trait, the same behavior different traits. Take a seemingly simple concept like "kindness." Kindness may be manifested in endlessly different ways—helping an old person across the street, consoling a crying child, giving milk to a hungry kitten, offering to prepare dinner for an ailing neighbor, contributing aid to the needy, etc. Furthermore, it is reflected in different behavior styles—a tone of voice, a facial expression, a gentle touch. It is evident that we are describing many different acts directed toward different objects expressed in different ways under different circumstances. Plainly, subtlety and sophistication of intellect are required to recognize the commonality among these very diverse acts.

If the concept expresses that element common to diverse and heterogeneous items, what the learning of a concept does is to alert us to that quality which diverse elements have in common, a commonality, which, in the absence of such learning, would otherwise have escaped our attention. In learning concepts, including trait concepts, then, the individual becomes alerted to aspects of human behavior, which, though seen, have not registered. Concepts, then, do not so much describe reality as represent an intellectual manipulation of reality. The concept selects, edits, highlights, and organizes reality in terms of a particular point of view.

When society, in the course of socialization, teaches the individual the meaning of trait concepts, it profoundly affects his self-concept by providing him with a point of view for viewing the self. Learning such ideas as smart or interesting or friendly alerts him to aspects of the self to which he might otherwise not have attended and causes him to see himself within these frameworks. The adolescent's superior ability to probe beneath the surface of things, to achieve a new intellectual synthesis of the materials of experience which impinge upon his senses, effects a profound change in his self-concept. The very structure of the self-concept is altered; older and younger children now see themselves within different intellectual frameworks, apply to themselves different categories of thought.

Finally, the emergence of the self-concept as a language of *interpersonal* traits is a product both of this higher order of conceptualization and of the advanced ability to see the world, including the self, from the viewpoint of others. This is apparent in the concept of the "looking glass self" which, according to Cooley (1912: 152), "has three principal elements: the imagination of our appearance to the other person; the imagination of his judgment of that appearance; and some sort of self-feeling, such as pride or mortification." The concept "well-liked," for example, includes all three elements: what we believe the other thinks of us; what we believe he feels about what he sees; and our own emotional response to his judgment. None of this is possible unless the individual is able to see himself as an object of observation by others. The ability to see oneself as well-liked, easy to get along with, or pleasant, then, is no mean feat; it is, on the contrary, the product of considerable intellectual and social development. It is only when the child is able to get beyond his own narrow view of the world that he can succeed in seeing himself as an object of observation of others, as one who *arouses in other people's minds a definite set of thoughts and feelings*. By taking the role of the other, the individual comes to *define* himself in terms of the reactions he arouses in the minds of others—as well-liked, popular, easy to get along with, and other looking-glass traits. Younger children are less likely to define themselves in terms of these interpersonal traits, we believe, because their social development has not reached the stage where they can see themselves as objects of observation of others.

In sum, development brings with it not only a change in self-concept components, but also a change in the *way* people think about themselves. At the heart of this change is the social factor of human communication. True communication involves the ability to adopt the viewpoint and perspective of the other. But this ability is not easily acquired; it requires a certain level of maturation. Once established, however, it makes possible a quality of social relationships previously impossible to achieve and at the same time effects a major transformation in the structure and content of the self-concept.

## NOTES

1. One nine-year old gave the following breathtaking recital of the things "not so good" about him: "Fighting, cussing, kicking, talking back to other people—to grown-ups. Busting windows. Setting a fire, stealing out of a store, cheating other little people,

like little children. Breaking in houses." This recital, we expect, was less a source of chagrin than an unbridled boast.

2. This classification, of course, lumps together a number of categories which, for other research purposes, may usefully be distinguished. By far the most elaborate and sophisticated scheme for the classification of self-concept components is that developed by Gordon (1968, 1974, 1976). See also Mulford and Salisbury (1964).

3. Although some will contend that this is a misunderstanding, a high level of detachment and objectification is necessary in the psychoanalytic process to hold up to conscious reflection those elements buried in the unconscious.

4. One reason for thinking that these differences are real, rather than merely artifacts of vocabulary, is that even a limited vocabulary will frequently suffice to communicate a concept if the idea is there in the first place. The child of 9–10 may not know the word "diligent," but he does know "works hard;" he may not know "punctual" but does know "on time" or "not late"; he may not know "conformist" but does know "always does what the other kids do"; he may not know "imitative" but does know "copycat"; and he may not know "vengeful" but does know "gets even." If the younger child is less likely to describe the self in trait terminology, it is less because he lacks the requisite vocabulary than because he lacks the concepts.

As a matter of fact, an examination of the responses even of the high school students reveals a rather small proportion answering in terms of specific adjectives; most often descriptive phrases are used. Instead of "secretive," they say: "I hide my feelings." Instead of self-controlled, "I'm calm," or "I don't lose my temper easily." "I don't get mad easily." Instead of frivolous or light-hearted, the adolescent sees himself as "a little clown because I like to joke around"; "I take things as a joke"; "Like to have fun." Instead of "independent" or "non-conformist," one hears "I don't follow crowds in fast dress; I wear what I want"; instead of frank and open, "I don't believe in hiding my feelings. If I have something to say, I say it." These descriptive phrases clearly show that the adolescent has mastered and can communicate the concept even though the adjectival term is alien to him.

Even if we select adjectives almost certainly familiar to both older and younger children, the older children are more likely to describe themselves in terms of these traits. As a simple experiment, we selected at random 98 high school and 101 elementary school children, focusing on the question "what kind of personality would you like to have, or what kind of person would you like to be, as an adult?" Without regard to order of response, we counted (among those who gave specific answers) how many of each said "friendly," "easy to get along with" (or "get along well"), and "liked" or "well-liked." The data show that the older children are far more likely to describe themselves using these *precise* words. Of 101 elementary school children, 9 children said "friendly," 2 said "easy to get along with," and 2 said "well-liked." Of 98 senior high school pupils, 20 said "friendly," 19 said "easy to get along with," and 16 said "well-liked." If the younger children are less likely to use these terms, it is surely not because they have not heard terms like "friendly," "easy to get along with," or "well-liked," but rather because they are less likely to think in terms of these concepts.

# 9

# Self-Concept Disturbance:
# Course and Development*

**W**HEREAS the previous chapter described changes in how young people *think* about themselves as they grow older, this chapter focuses on changes in their *feelings* about themselves. Among the most widely accepted ideas in the behavioral sciences is the theory that adolescence is a period of disturbance for the child's self. Hall (1904) originally characterized the age as one of "storm and stress." Erikson (1959) views it as a time of identity crisis in which the child struggles for a stable sense of self. Psychoanalytic theory postulates that the sexual desires of puberty spark a resurgence of oedipal conflicts for the boy and pre-oedipal pressures for the girl (Blos, 1962; A. Freud, 1958). The physiological changes of puberty and the increase in sexual desire challenge the child's view of himself in fundamental ways. Both his body-image and his self-concept change radically.

Sociologists (Davis, 1944) traditionally characterize adolescence as a period of physical maturity and social immaturity.[1] Because of the complexity of the present social system, the child reaches physical adulthood before he is capable of functioning well in adult social roles. Adolescence becomes an extremely difficult period because the new physical capabilities and new social pressures to become inde-

* This chapter was written in collaboration with Roberta G. Simmons and Florence R. Rosenberg.

pendent coincide with many impediments to actual independence, power, and sexual freedom.

The resulting status ambiguities (that is, the unclear social definitions and expectations), have been seen as engendering a corresponding ambiguity of self-definition. In addition, the need to make major decisions about future adult roles on the basis of what he is like at present further heightens the adolescent's self-awareness and self-uncertainty (Erikson, 1959).

With all these physical, emotional, and social changes, it is small wonder that many social theorists assume that this is a difficult period for the child. Some investigators, however (Offer, 1969; Grinker *et al.*, 1962; Douvan and Adelson 1966) have questioned whether there is an adolescent crisis. Several fundamental questions on the self-concept remain to be resolved.

First, do data support the belief that the adolescent's self-concept differs from that of younger children? If so, could one term this difference a "disturbance," that is, a change which would cause the child some discomfort or unhappiness?

Second, if there is an adolescent self-concept disturbance, when does it begin? Resolution of this question is crucial to the evaluation of certain theoretical notions. Erikson (1959) tells us that the adolescent must deal with the issues of a career decision and the establishment of his own family. Although these concerns may be salient to the 18 or 19 year old, they do not concern the 12 year old equally. Conversely, it is the younger adolescent who is confronted with the changes in physical characteristics accompanying puberty. This chapter tries to specify the time of onset of adolescent self-concept disturbance.

Third, if there is an adolescent self-concept disturbance, what is the course of its development? Do the problems appearing at the time it is precipitated continue to grow? Do they level off at a higher plane? Or do they decline as the adolescent learns to cope with them?

Finally, if there is an adolescent disturbance, what triggers it? Typically, the onset of puberty is viewed as the cause. The question is: can one uncover patterned social experiences which also play a role?

## DIMENSIONS AND MEASURES

In this chapter, we shall examine the relationship of age to the following six variables: global self-esteem, stability of the self-concept, self-consciousness, valued specific components, perceived self, and depressive affect. Since self-esteem and stability have already been described in chapter 4,[2] the other four measures will be briefly discussed here.

Self-consciousness primarily has reference to the *salience* of the self, that is, the degree to which the self as an object is prominent in one's thoughts, is in the forefront of attention. Maslow (1956: 244) observes that "another characteristic of healthy people . . . is that of spontaneity. Spontaneity means that the behavior of this individual tends to be natural, easy, unself-consciousness, and to flow automatically without design and without intent." The self is chiefly salient when it is problematic; people who accept themselves, or feel comfortable with themselves, are not likely to keep the self in the forefront of attention. According to Murphy (1947), "If the self fulfills one's expectations, it may remain a peripheral experience; if it falls short it may become prominent."

Self-consciousness was measured in the Baltimore study by a 7-item Guttman scale with satisfactory internal reliability.[3] (Example: "Let's say some grownup or adult visitor came into class and the teacher wanted them to know who you were, so she asked you to stand up and tell them a little about yourself. Would you like this, would you not like it, or wouldn't you care?").

The subject's assessment of the following eight specific dispositions will also be examined: smart, good-looking, truthful and honest, good at sports, well-behaved, work hard in school, helpful, and good at making jokes. But since, as we indicated in chapter 2, a negative assessment will make a difference only if the individual cares about the particular component, we shall consider only those subjects who highly value each component.

The final aspect of the self-concept to be considered is the "perceived self"—what the individual believes others think of him. In the Baltimore study (as noted in chapter 3), the pupils were asked to indicate whether or not their mothers, fathers teachers, kids in their class, boys and girls thought well of them; and whether, if parents, teachers, and peers were to describe them to other people, these remarks would be favorable or unfavorable.[4]

Finally, in order to learn whether or not there is a parallel between self-concept disturbance and other types of emotional disturbance, we shall examine "depressive affect." This variable was measured in Baltimore by a 6-item Guttman scale with satisfactory internal reliability.[5] (Examples: "A kid told me: 'Other kids seem happier than I.' Is this true for you or not true for you?" "Would you say that most of the time you are very cheerful, pretty cheerful, not very cheerful, or not cheerful at all?").

## Self-Concept Disturbance

Table 9–1 shows the relationship of age to global self-esteem, stability of self-concept, self-consciousness, perceived self, and depressive affect. Table 9–2 indicates how younger and older children rate themselves on the eight specific qualities mentioned above (based only on those respondents who said they cared "very much" about the particular characteristic). These data rather clearly indicate that the emergence of self-concept problems in adolescence is no·myth, and that these problems occur *early* in adolescence. In general, self-concept disturbance appears to be much greater in the 12–14 age group than in the 8–11 age group.

In contrast to younger children, the early adolescents (12–14 year olds) show a higher level of self-consciousness, greater instability of the self-concept, slightly lower global self-esteem, and less favorable judgments of valued self-components, and (with the exception of opposite sex judgments) less favorable perceived selves. The assumption that such changes are likely to be disturbing is consistent with the fact that early adolescents also show a higher level of depressive affect than do the younger children (Table 9–1).

Whereas the early adolescents show a heightened self-consciousness and a greater degree of instability of the self-picture, this self-consciousness and instability levels off *in later adolescence*. Generally, in late adolescence, the subjects manifest greater self-consciousness, greater instability, lower ratings on valued self-concept components (Table 9–2), and lower perceived selves than do the 8–11 year old children. Only in the case of global self-esteem is there an improvement in later adolescence marked enough for the youngsters from age 15 up to score more favorably than the 8–11 year olds. The older adolescents show higher global self-esteem than both the young children and the early adolescents.[6]

## TABLE 9-1.

*Self-Concept Dimensions and Depressive Affect, by Age*

| | 8-11<br>(N = 786) | | 12-14<br>(N = 637) | | 15+<br>(N = 502) |
|---|---|---|---|---|---|
| | | | **Age** | | |
| *Self-concept dimensions* | | | | | |
| *High self-consciousness | 18% | * | 35% | | 33% |
| *High instability of self-concept | 25% | * | 37% | | 35% |
| *Low global self-esteem | 25% | | 29% | * | 22% |
| *Negative perceived self* | | | | | |
| *Parents negative† | 51% | * | 64% | | 64% |
| *Teachers negative† | 83% | * | 93% | | 94% |
| Opposite sex peers negative‡ | 84% | | 87% | | 84% |
| *Same sex peers negative | 51% | * | 71% | * | 86% |
| *Depressive affect* | | | | | |
| *High depressive affect | 18% | * | 31% | | 34% |

*Asterisk before variable indicates significance for total $\chi^2$ test. Asterisk between adjacent age groups indicates significance of difference between proportions. Because of nonindependence of comparisons, tests between adjacent age groups are not entirely appropriate, but are included here as an indication of how seriously to take the observed differences (see Blalock, 1972: 328-34 for a discussion of this issue in the case of analysis of variance).

†Think child is not at all, a little, or pretty nice.

‡Like child not much or pretty much.

## TABLE 9-2.

*Proportion Rating Selves Very Favorably on Each Characteristic Among Those Who Care "Very Much" About That Characteristic, by Age*

| | **Age** | | |
|---|---|---|---|
| | 8-11 | 12-14 | 15+ |
| *Rate self very favorably on . . .* | | | |
| *Smart | 26% | 9% | 5% |
| N = 100% | (547) | (366) | (244) |
| *Good-looking | 20% * | 13% * | 6% |
| N = 100% | (258) | (197) | (121) |
| *Truthful or honest | 54% * | 38% | 38% |
| N = 100% | (527) | (424) | (320) |
| *Good at sports | 50% | 46% | 42% |
| N = 100% | (339) | (266) | (163) |
| *Well-behaved | 46% * | 31% * | 39% |
| N = 100% | (474) | (332) | (239) |
| *Work hard in school | 71% * | 50% * | 39% |
| N = 100% | (494) | (373) | (231) |
| *Helpful | 60% * | 46% | 46% |
| N = 100% | (506) | (329) | (225) |
| *Good at making jokes | 49% | 40% | 46% |
| N = 100% | (151) | (53) | (28) |

*See note, Table 9-1.

To summarize, the results show a general pattern of self-concept disturbance in *early* adolescence. Compared to the younger children, the early adolescent has become distinctly more self-conscious; his picture of himself has become more shaky and unstable; his global self-esteem has declined slightly; his attitude toward a number of the specific characteristics which he highly values has become less positive; and he believes others view him less favorably. In view of all these disturbing changes in the self-picture, it is hardly surprising that our data show early adolescents to be significantly more likely to be depressed.

The course of self-concept development after the 12–14 age period is also interesting. In general, the differences between early and late adolescence are not large. Global self-esteem improves somewhat and same sex perceived selves decline, but in general early and late adolescents show similar results. The chief difference is almost always between the 8–11 year old children and the 12–14 year old children. It appears that it is at this age that changes in these self-concept dimensions appear and that they are of a distressing variety.

## Onset of the Disturbance

The data thus show distinct changes in the self-concept from childhood to early adolescence. The question is: can one be more specific about this change? Does the change occur gradually or suddenly?

As Figure 9–1 shows, there appears to be a noticeable difference between children who are 11 years old and those who are 12 years old. (It should be noted that the 11-year old group includes children from age 11 years no months to 11 years 11 months, while the 12-year old group covers those from 12 years no months to 12 years 11 months.) Self-consciousness, instability of the self-concept, low global self-esteem, high depression, low valued specific self-traits, and negative perceived selves all rise relatively sharply among the 12-year olds as compared to the 11-year olds, although in most cases some rise has begun earlier, particularly the year before. This movement from the 11-year old group to the 12-year old group is the only one-year period in which the children show an increase of disturbance on all these measures. In fact, on all four measures it is the largest yearly increase in disturbance to occur up to that age. In the case of three out of four measures, it is the largest increase to occur between any two ages.

In terms of almost all the dimensions considered here, disturbance continues to increase after 12, but in most cases the highest point of disturbance is either at age 12, 13, or 14. In fact, stability of the self-concept and global self-esteem seem to improve after this point, particularly in very late adolescence, while disturbances in self-consciousness, specific valued self-traits, and perceived self seem to level off and remain at the early adolescent levels.

Global self-esteem development deserves special mention. We noted earlier that the self-esteem level of the 12–14 year group was only slightly lower than that of the 8–11 group. But this finding conceals an important change: the sudden dramatic decline in self-esteem among the 12-year olds (Figure 9–1: compare 11 and 12-year olds). But during the following year, when the children reach 13, global self-esteem rapidly returns to its earlier level and continues to rise in later adolescence. In our later discussion, we shall see some possible reasons for this drastic but temporary shift.

In sum, the data suggest that during their twelfth year (that is, between their twelfth and thirteenth birthdays), children tend to experience marked increases in self-concept disturbance. With regard to some dimensions, this relatively sharp increase continues through the twelfth and thirteenth years. It is relevant to note that early adolescence is also characterized by a corresponding increase in feelings of depression, though this rise has clearly begun earlier. After age 13, there is again a general leveling off. The conclusion appears inescapable: early adolescence—the twelfth and thirteenth years—is characterized by a greater rise in self-concept disturbance than that which occurs at any other point in time.

ENVIRONMENTAL CONTEXT

The rise in self-concept disturbance at some time after the twelfth birthday would obviously suggest that a major determinant of this stress is the onset of puberty—that biological forces and changes in physical characteristics are essentially responsible for the disturbance. The question is: are there any factors in the social environment which may also contribute to these changes?

There is one important environmental change which occurs for most children at this time. Children generally begin their last year of

**FIGURE 9-1.**

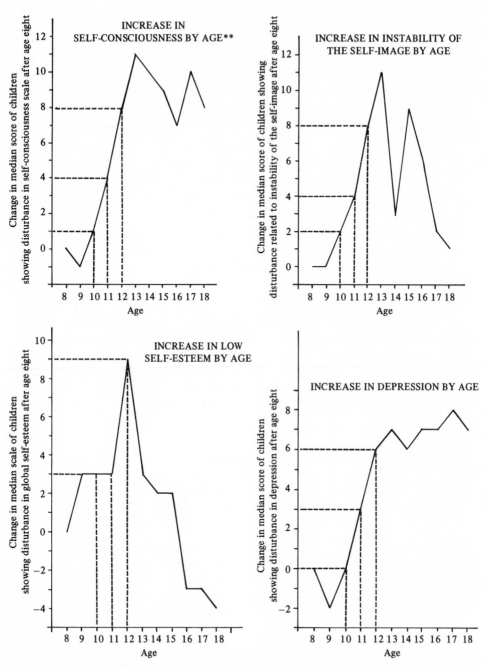

*For each scale, the median score of the eight-year old group is subtracted from the median score of each subsequent age group. If the graph line rises, then disturbance along the dimension is said to increase. The points above "0" indicate a higher level of disturbance after age 8, while the points below "0" indicate a lower level.

**The values at age 10, 11 and 12 are indicated by dotted lines. In the sample there are 98 eight-year olds, 225 nine-year olds, 263 ten-year olds, 233 eleven-year olds, 237 twelve-year olds, 213 thirteen-year olds, 199 fourteen-year olds, 150 fifteen-year olds, 162 sixteen-year olds, 130 seventeen-year olds and 56 eighteen-year olds.

elementary school (the sixth grade) when they are 11 and begin the first year of junior high school (the seventh grade) when they are 12. Does the movement into junior high school itself appear to contribute to the increase in self-concept disturbance?

Obviously, one cannot examine the effects of change in environment by directly comparing sixth and seventh graders since one does not know whether such differences are due to the fact that the seventh graders are in junior high school or simply that they are older. It is, however, possible to disentangle the effects of age maturation and school contexts by comparing children of comparable ages. By the spring of the school year, when our data were collected, there was an appreciable number of 12-year olds in both the sixth and seventh grades. If the junior high school experience were particularly stressful for the child, then the 12-year olds in junior high school should show greater disturbance of their self-concepts than the 12-year olds in elementary school.

Table 9–3 provides dramatic support of this hypothesis. The 12-year olds in junior high school have lower global self-esteem, higher self-consciousness, and greater instability of the self-concept than their age-peers in elementary school. For example, 41 percent of the 12-year olds in junior high school have low global self-esteem in contrast to only 22 percent of those in elementary school; 43 percent of the former manifest high self-consciousness in comparison to only 27

**TABLE 9-3.**

*Disturbance of the Self-Concept by School Context,*
*Among Twelve-Year Old Children*

|  | Twelve-Year Old Children | | |
|---|---|---|---|
|  | In elementary school | | In junior high school |
| *Self-concept disturbance* | | | |
| Low global self-esteem | 22% | * | 41% |
| N = 100% | (167) | | (59) |
| High self-consciousness | 27% | * | 43% |
| N = 100% | (172) | | (61) |
| High instability of self-concept | 30% | * | 53% |
| N = 100% | (158) | | (60) |

*See footnote 7, chapter 3, p. 98.

percent of the latter; and 53 percent have relatively unstable self-concepts compared to 30 percent in elementary school.

These findings afford a vivid illustration of the way a social context can affect individual personality. But the possibility must be considered that the 12-year olds still in sixth grade differ in other ways from the 12-year olds in seventh grade. We do find that the 12-year olds still in the sixth grade are more likely to have lower school marks and to be from the lower social classes, but there is certainly nothing in this to *enhance* their self-concepts. In any case, when we control on these factors using test factor standardization (Rosenberg, 1962b), we find that none of the original differences involving global self-esteem, specific self-components, self-consciousness, perceived selves, or stability of the self-concept can be explained by any of these variables. Even when standardized on class or marks in school, all differences between elementary school and junior high school 12-year olds remain essentially unchanged. Furthermore, Table 9-4, dealing with the three global measures of self-esteem, self-consciousness, and instability, shows that these findings generally hold for blacks as well as whites, for middle-class as well as working-class respondents, and for students with high grades as well as those with low grades.

The above argument assumes that once these variables are controlled the only remaining difference between these two types of 12-year olds is the school they attend. However, one other possibility involves the relative ages of these two groups to others in their classes. The sixth grade 12-year olds are among the oldest and biggest children in their class whereas the seventh grade 12-year olds are among the youngest and least physically mature children in their class. It is possible that the self-concepts of the sixth grade 12-year olds benefit from their relative advantage whereas the self-concepts of their seventh-grade age peers suffer from their age rank in their group. If this hypothesis explains the findings, then the sixth grade 12-year olds should have more positive self-concepts than the younger children in their classes whereas the seventh-grade 12-year olds should show more disturbed self-concepts than the older children in their grade.

Yet Table 9-5 shows there is virtually no difference between the self-concept ratings of 11 and 12-year olds in the sixth grade, nor is there a difference between the self-concepts of the 12 and 13-year olds in the seventh grade.

Thus, the transition from elementary to junior high schools seems

## TABLE 9-4.

Disturbance of the Self-Concept Among Twelve-Year Olds in the Sixth or Seventh Grade, by Race, Social Class, and Marks in School

| | Race | | | | Social Class | | | | Marks in School | | | |
| | Blacks | | Whites | | Middle Class | | Working Class | | A's and B's | | C's and Below | |
| | 6th grade | 7th grade | 6th grade | 7th grade | 6th grade | 7th grade | 6th grade | 7th grade | 6th grade | 7th grade | 6th grade | 7th grade |
|---|---|---|---|---|---|---|---|---|---|---|---|---|
| **Self-Concept disturbance** | | | | | | | | | | | | |
| Low global self-esteem | 18% | 33% | 30% | 47% | 14% * | 44% | 21% | 36% | 20% | 37% | 23% | 42% |
| N = 100% | (106) | (27) | (61) | (32) | (21) | (16) | (119) | (39) | (49) | (30) | (104) | (24) |
| High self-consciousness | 28% | 32% | 27% * | 52% | 24% | 53% | 29% | 38% | 22% | 37% | 29% * | 58% |
| N = 100% | (109) | (28) | (62) | (33) | (21) | (17) | (122) | (40) | (50) | (30) | (106) | (26) |
| High instability of self-concept | 31% | 50% | 26% * | 56% | 35% * | 71% | 27% * | 44% | 30% | 52% | 28% * | 54% |
| N = 100% | (104) | (28) | (53) | (32) | (20) | (17) | (112) | (39) | (47) | (29) | (96) | (26) |

*See footnote 7, chapter 3, p. 98.

**TABLE 9-5.**

*Disturbance of the Self-Concept by Age, by Grade in School*

| | Grade in School | | | |
| --- | --- | --- | --- | --- |
| | Sixth Grade | | Seventh Grade | |
| | Age 11 | Age 12 | Age 12 | Age 13 |
| *Self-concept disturbance* | | | | |
| Low global self-esteem | 29% | 24% | 40% | 36% |
| N = 100% | (73) | (106) | (58) | (101) |
| High self-consciousness | 22% | 27% | 42% | 46% |
| N = 100% | (76) | (106) | (60) | (101) |
| High instability of self-concept | 30% | 33% | 53% | 46% |
| N = 100% | (76) | (102) | (59) | (92) |

to represent a significant stress along several dimensions of the child's self-concept; on the other hand, aging from 11–12 or from 12–13 does not in itself appear stressful. Within the same school class, age makes little difference; but within the same age group, school class makes a great difference.

Furthermore, it is not just change *per se*, but change from elementary to junior high school, that is associated with disturbance. We find, for example, that the transition from junior to senior high school does *not* show a parallel effect on the self-concept; 15-year olds in senior high school do not show more disturbed self-concepts than 15-year olds in junior high school. It is thus the specific transition to a *junior high school* that generates the disturbance.

## Discussion

According to these data, self-concept disturbance appears to reach its peak in early adolescence, in some respects declining in later adolescence, but in most respects persisting. In many areas, a particular rise in disturbance appears to occur when the child is 12, that is, between the twelfth and thirteenth birthdays. The rise frequently has started a year before, and it may continue to rise for the next year or so; often, however, it does not appear to increase much, if at all, after age 13 or 14.

During early adolescence, in comparison to the childhood years of 8–11, the children exhibit heightened self-consciousness, greater instability of the self-concept, slightly lower global self-esteem, lower opinions of themselves with regard to certain qualities they value, and a reduced conviction that their parents, teachers, and peers of the same sex hold favorable opinions of them. It is thus interesting to note that they are also more likely to show high depressive affect.

These data agree with the findings of Offer (1969, Ch. 11), who studied a somewhat older adolescent group (14–18), and who reports that both parents and adolescents agreed that the greatest amount of "turmoil" in their lives occurred between ages 12 to 14. The specific finding that instability of the self-concept increases during adolescence might appear to support Erikson's (1959) views on adolescent problems of ego-identity. However, Erikson seems to see the ego-identity crisis as occurring in late adolescence whereas our data indicate a rise in instability of the self-concept during the early adolescent years.

What is there about early adolescence, primarily the 12–13 age period, that is so crucial from the viewpoint of self-feelings? At this time of life, three events are occurring that overturn the child's world and bring the self to the forefront of attention in a new and dramatic way. The first is biological, the second, environmental, the third, interactional.

The onset of puberty, with the startling surge of sexual desire and its accompanying physical changes, is a direct and serious challenge to the taken-for-granted self (Simmons, Blyth, Brown, and Bush, 1977). So long as the child has worked out a *modus vivendi* based on a set of implicit and unreflective self-assumptions, the self can remain in the background of awareness, accepted for what it is. But if conspicuous and overt changes in the actual self occur, particularly to those disposed to conceptualize the self as a social exterior, then the self moves to the forefront of attention. Physical features are a case in point. For most of us, our bodily concepts represent an image we carry about with us in the middle ground of consciousness. But if we were to wake up one morning and find ourselves six inches taller, 20 pounds lighter, and having different hair color, head shape, and facial structure, we would turn our attention squarely to these facts. The same would be true if we were suddenly to experience powerful and hitherto unknown endocrinological sensations; the entire structure of the taken-for-granted self would come under challenge. Such questions as "who am I" or "what am I" would surge to the fore.

Yet this is only an exaggeration of what happens in puberty. Puberty, involving the appearance of sexual urges with accompanying physical changes, is a direct and dramatic challenge to the taken-for-granted self. The individual is no longer certain of who he is, what he is, and what he feels. It is thus no accident that the most marked changes of early adolescence are not a decline in global self-esteem—although that also happens—but a sharp rise in both self-consciousness and in self-concept instability.

Coincident with this experience, though independent of it, is an important environmental change in the lives of children—the shift from elementary to junior high school. Selves exist in certain contexts, including the expectations of familiar others, and as long as these contexts remain fixed, the self can be taken-for-granted and remain stable. It is only when we are uncertain about others' expectations of us or our expectations of ourselves that the self comes sharply to the fore.

Yet this is what the transition from elementary to junior high school represents. It is, indeed, the shift from the sixth to the seventh, rather than prior or subsequent, grades that is particularly associated with heightened self-consciousness and increased instability. In this environment, the individual is faced with a new coterie of peers and a multiplicity of teachers with whom no firm set of mutual expectations has been established. Many new intellectual challenges, such as those involved in the mastery of mathematics or foreign languages, must be confronted; the first decisions setting him on the path to a career (whether to enter an academic, vocational, commercial, or agricultural track) must be made. All this calls for probing self-assessment. The old taken-for-granted self is undermined. The self-concept becomes problematic, characterized by high self-consciousness, instability, and a questioning of self-worth.

With advancing age and consistency of circumstance, the shock wears off, new expectations are built up, and a new equilibrium is reached. The adolescent learns what others expect of him, establishes a more stable view of his strengths and weaknesses, gains a new appreciation of the self. To be sure, he can never return to the unreflective self of childhood, especially since, as Erikson (1959) has stressed, he must now think about the self in relation to the adult tasks of entering the world of work and of establishing a family.

Nevertheless, the immediate context is sufficiently familiar to reestablish the stable, unproblematic self-concept.

But probably the most important factor underlying the sharp rise in self-consciousness and instability in early adolescence is the decisive victory over egocentrism (Piaget, 1948) which permits, for the first time, full-fledged communication. It is at this point that the child finally succeeds in getting beyond his own point of view and of seeing himself from the viewpoint of others.

As we pointed out in the previous chapter, when the adolescent comes to conceptualize the self in terms of interpersonal traits (as the younger child does not), he is defining himself as an object of observation, as a person who arouses characteristic reactions in the minds of other people. But this important development, which makes possible a new quality of human communication, also generates an enhanced self-consciousness. The individual becomes keenly aware of himself as an object of observation of others. Whereas in childhood the responses of others were taken for granted, in early adolescence the youngster becomes alert and sensitive to what others think of him—nervous when others watch him work, uncomfortable if he must perform before a group, concerned with the impression he makes on others. A sharp rise in self-consciousness is inevitable.

Uncertainty and instability of the self-concept follow closely on this development. A major reason is that the matter of entering into the mind of another and knowing what he is thinking is a mysterious and elusive business. Because no one can ever enter the mind of another with perfect accuracy, the veil is ultimately impenetrable. So long as the young child makes no attempt to see himself from the viewpoint of others, he can operate confidently according to a set of established assumptions of what he is like. But once he attempts to see himself through others' eyes, a whole new order of complexity is introduced. First, there are many "others" who enter his life, each viewing him from his own particular perspective. Second, their attitudes toward him are wrapped in uncertainty, thus providing a shaky foundation for a self-concept. One's sense of worth, in general or in specific respects, formerly accepted unreflectively, now comes under scrutiny.

Once the individual seriously attempts to fathom what others think of him, the conclusions that emerge are by no means palatable. This is evident in the striking decline in the "perceived self" in early adolescence. In the early years, the child has accepted the idea that others think well of him and has given the matter little thought. In early adolescence, when he is able to get outside his own viewpoint and see matters, including himself, from the perspective of others, he recognizes the complexity of others' attitudes toward him; he becomes aware of both their qualms and their sources of appreciation. Except with regard to inferences about the attitudes towards him of the other sex (which has radically different significance in childhood and adolescence), his views of others' attitudes toward him decline sharply.

On the basis of these data, we are driven to conclude that the degree of negative self-concept change may be of sufficient magnitude to warrant the term "crisis"; at the least, it is highly disturbing. Although this disturbance will start earlier and continue later, it is likely to center on the age period 12–13. The child becomes much more aware of what others think of him, much less sure of what he is like, much less convinced that others approve of him, and somewhat more self-critical in general and with reference to specific valued components. The developments which stimulated these changes, naturally, do not disappear overnight (sexual urges increase, radical bodily changes continue, friendships become increasingly intense, concern with the attitudes of others remains high, situations change) but in later adolescence the individual succeeds in establishing a new *modus vivendi* in relation to these problems. Although his global (but not specific) self-esteem again rises, it probably now rests on new and fundamentally different foundations, based on a self-aware assessment of his qualities and of the attitudes of others toward the self. Thus, even though matters improve in some degree in later adolescence, the individual never again (at least during pre-adulthood) attains the unreflective and confident self-acceptance of childhood.

## NOTES

1. See Gordon (1971), Bakan (1971) and Kohlberg and Gilligan (1971) for discussions of adolescence as a social phenomenon.
2. Details concerning these measures appear in Appendix A-2 and A-4.
3. The Coefficient of Reproducibility is 89.4 percent, the Coefficient of Scalability is 62.5 percent, and the percent improvement is 17.8 percent. A description of the items and scoring procedure is presented in Appendix A-6.

4. The perceived self-concept indicators are presented in Appendix B-1.

5. Coefficient of Reproducibility = 92.2 percent, Coefficient of Scalability = 69.5 percent, percent improvement = 17.8 percent. The six items in this scale and the scoring procedures are presented in Appendix A-5.

6. Earlier studies (Engel, 1959; Piers and Harris, 1964) have also shown an increase in self-esteem among senior high school students; furthermore, one of these (Piers and Harris, 1964) demonstrated a decline in self-esteem in early adolescence as compared to childhood.

# 10

## The Locus
## of Self-Knowledge

THE PREVIOUS two chapters have shown how both thoughts
and feelings about the self-concept change as children grow older. But
there are also certain interesting changes in the individual's view of
*knowledge* about the self. In examining this question, we shall adopt
the view that knowledge about the self can be likened to knowledge
about anything else. In thinking of any sort of knowlege, we tend to
operate on the assumption that there is a truth and that someone
knows it—or is closer to it than others. We accept the encyclopedia's
statement of Napoleon's birth date, the mathematics professor's answer to a math problem, the minister's version of moral truth, the
lawyer's version of legal principle. But whose version of the truth about
the self do we accept? This truth, having very special importance for
us, may be within or without—internal to the self or external to it.

The key question of this chapter is the following: who, in our
view, really knows best what we are like? Where, in other words, does
the locus of self-knowledge lie? At first glance, the answer would
seem self-evident since there is, after all, only one final authority
on the subject—ourselves. However more knowledgeable others may

be about anything else in the world, if there is one subject on which we are the world's unquestioned experts, on which we are uniquely qualified to pass judgment, it is surely ourselves. To claim that someone else knows what we are really like is, in effect, to conclude that his opinion of us agrees with our own.

But outstanding authority is not sole authority. We can surely acknowledge that certain others understand us to a greater or lesser extent. The question is: can faith in others' knowledge be so strong that we even accept the idea that they know us better than we know ourselves?

The degree of faith in others' judgments of us has obvious self-esteem implications. As we showed in the earlier discussion of significant others (chapter 3), there are many other people who hold views of us; some of these people's opinions are more important (valuation) or are more highly trusted (credibility) than others. Furthermore, there are many different components of the self; hence, a person may be acknowledged as an expert with reference to one component of the self, but his views given little credence with reference to another. Hence, in this chapter, we shall focus both on the *locus* of self knowledge (*who* knows what we are like) and the locus of *self*-knowledge (*what* about us do they know).

Among children of different ages, is the *locus* of self-knowledge to be found in adults (chiefly parents, teachers, and other adult relatives); peers (chiefly close friends, classmates or siblings); or the child himself? Two aspects of the self will be considered. The first refers to relatively exterior specific self-concept components. These are selected characteristics whose behavioral expressions or manifestations are relatively susceptible to public perusal, scrutiny, and evaluation. Specifically, the three traits to be considered are "smart," "good-looking," and "good." Attractiveness is obviously an overt and visible characteristic. Intelligence and morality, of course, are underlying dispositions; nevertheless, they are rather clearly reflected in behavior or behavioral outcomes, and frequently elicit explicit responses from others. For example, our parents tell us explicitly that we have been bad, and our report cards explicitly evaluate our intelligence.

The second aspect of the self-concept is more closely related to the individual's psychological interior. It is concerned with the covert, invisible, internal aspects of the self—with the individual's phenomenal field. Specifically, it refers to whether certain others know what we are like "deep down inside."

## Locus of Exterior Knowledge of the Self

That it is possible to believe that others know certain things about us better than we ourselves do has been strikingly documented in John Campbell's (1967, 1975) studies of perceptions of illness among children. For example, how does a young child decide whether he is sick? Is this decision based on direct experience, that is, does he feel hot, tired, nauseous, in pain, or dizzy? No, the young child concludes he is sick chiefly because his mother tells him so. He relies not on the internal and otherwise inaccessible experience of illness, but on the external authority of the mother.

But what about judgments of morality, intelligence, and aesthetics? Where does the final authority on these matters rest? In order to gain some purchase on this issue, we asked our Baltimore subjects the following questions:

1. If I asked you and your mother how smart you were, and you said one thing and she said another, who would be right—you or your mother?
2. If I asked you and your mother how good your were, and you said one thing and she said another, who would be right—you or your mother?
3. If I asked you and your father how good-looking you were, and you said one thing and he said another, who would be right—you or your father?
4. Who knows better whether you are good—you or your mother?

The younger children, it turns out, appear to have even greater faith in their parents' knowledge of their moral, intellectual, and aesthetic characteristics than in their own. In the 8–11 age group, between 67 and 78 percent held that, in the event of a disagreement about one of his characteristics between himself and a parent, *the parent* would be right. Among respondents 15 or older, on the other hand, between 50 and 68 percent said that *they* would be right. The older the child, then, the more firmly is truth about the self anchored within the self.

In order to bring the results into sharper focus, these four items were combined to form a "locus of exterior self-knowledge" score. (Because one of the items dealt with fathers, children without fathers are excluded from the remainder of these analyses.) As Table 10–1 shows, the developmental changes are striking. Almost half of the older children, but less than one-sixth of the younger children, placed

**\*TABLE 10-1.**

*Locus of Exterior Self-Knowledge, by Age*

|  | Age | | |
| --- | --- | --- | --- |
|  | 8-11 | 12-14 | 15+ |
| *Exterior self-knowledge centered in . . .* | | | |
| Self 1 | 15% | 38% | 48% |
| 2 | 15 | 20 | 25 |
| 3 | 28 | 23 | 21 |
| Adult 4 | 42 | 19 | 6 |
| N = 100% | (494) | (386) | (284) |

\*See footnote 7, chapter 3, p. 98.

the locus of exterior self-knowledge within the self. Conversely, 70 percent of the younger, compared to 27 percent of the older, attributed the deeper wisdom to the adult (gamma = .4613).

With regard to exterior components, then, it is possible to believe that *others may know us better than we know ourselves.* They can judge how good, smart or handsome we are by observing how we look, act, or speak. Since these characteristics are relatively visible and public, it is meaningful to pit others' judgments on these matters against our own, and, given sufficient respect, to bow to their superior wisdom. The younger children are vastly more disposed to do so than the older ones.

## Locus of Interior Knowledge of the Self

When we turn to interior self-elements, we are dealing with components which are less accessible to the direct perusal of others. These represent the more intimate, secret aspects of our inner lives, aspects which we share with few or none—our dreams, ideals, worries, thoughts, feelings, and aspirations. Few people are either able to, or are permitted to, enter this deepest realm of the self. Furthermore, even under the most propitious circumstances, they can only know it at second hand, for it is removed from their direct experience.

With regard to this interior self, then, two questions will be considered: (1) does the child believe that adults or peers know it better? (2) does the child believe that adults know it better than he himself does?

ADULTS AND PEERS

It will be recalled from the discussion of significant others that the Baltimore school pupils were asked the following: "Who do you feel really understands you best? I mean, who knows best what you really feel and think deep down inside?" Subjects were then asked whether their mothers, fathers, any siblings, any friends, and any teachers knew what they were like deep down inside.

The data show that the younger the child, the greater the likelihood that he thinks some adult (mother, father, teachers) knows what he really feels and thinks "deep down inside." Indeed, the number of younger children (8–11) believing that adults are aware of their deepest inner thoughts and feelings is astonishing—79 percent attribute such knowledge to their mothers, 63 percent to their fathers, and 47 percent to their teachers. Among the older children, (15+) the corresponding figures are 53 percent, 31 percent, and 23 percent. Furthermore, when we combine the three questions dealing with faith in mother's, father's, and teacher's knowledge to form an index of "adult locus of inner self-knowledge," (Table 10–2) the results are still more striking: among the younger children, 71 percent exhibited such high faith, (score 1–2) whereas among the older, only 37 percent had that much confidence in adult insight (gamma =

**\*TABLE 10-2.**

*Faith in Adult Inner Self-Knowledge, by Age*

|  | Age | | |
| --- | --- | --- | --- |
|  | **8-11** | **12-14** | **15+** |
| *Faith in adult knowledge* | | | |
| High | 71% | 50% | 37% |
| Medium | 16 | 25 | 28 |
| Low | 13 | 25 | 35 |
| N = 100% | (488) | (400) | (306) |

*See footnote 7, chapter 3, p. 98.

.3813). To the adult, perhaps, the inner workings of the childish mind may be one of the great mysteries of the universe, but to the child himself he is, relatively speaking, an open book.

When we turn to peers' insight into what we "really feel and think deep down inside," the results are reversed: older children are more likely than the younger to believe that their friends and siblings know best what they are really like deep down inside. Combining these two indicators of "peer locus of inner self-knowledge" shows that 34 percent of the older, but 17 percent of the younger, children believed their peers knew what they really felt and thought. (Table 10–3).

**\*TABLE 10-3.**

*Faith in Peer Inner Self-Knowledge, by Age*

|  | Age | | |
|---|---|---|---|
|  | 8-11 | 12-14 | 15+ |
| *Faith in peer knowledge* |  |  |  |
| High | 17% | 23% | 34% |
| Medium | 33 | 40 | 36 |
| Low | 50 | 37 | 30 |
| N = 100% | (487) | (393) | (314) |

*See footnote 7, chapter 3, p. 98.

Finally, when asked who has the *greatest* knowledge of our most intimate feelings, the unilateral or vertical respect of the younger and the mutual or horizontal respect of the older become particularly clear. Asked to pick the *one* person (himself excepted) who knows best what he really thinks and feels deep down inside, 84 percent of the younger children but 52 percent of the older chose one or both parents (chiefly the mother). Conversely, only 7 percent of the younger children but 37 percent of the older ones chose a friend or sibling (Table 10–4).

The declining faith in adult insight does not mean that older children lack respect for parents. Table 10–4 indicates that older children have greater faith in adults (usually the mother) than in peers; this faith, however, is not nearly as strong as that of the younger children.

**\*TABLE 10-4.**

*Person Who Knows Best What Subject is Like*
*Deep Down Inside, by Age*

|  | Age | | |
|---|---|---|---|
|  | 8-11 | 12-14 | 15+ |
| *Knows you best deep down* | | | |
| Mother | 59% | 53% | 40% |
| Father | 10 | 10 | 9 |
| Both Parents | 15 | 8 | 3 |
| Total (parents) | *84%* | *71%* | *52%* |
| Teacher | — | — | 1 |
| Sibling | 2 | 9 | 12 |
| Friend | 5 | 12 | 25 |
| Other | 5 | 4 | 4 |
| Total (other person) | *12%* | *25%* | *42%* |
| No one | 3 | 4 | 6 |
| N = 100% | (477) | (395) | (283) |

*See footnote 7, chapter 3, p. 98.

ADULTS VERSUS SELF

We now press the idea of faith in an external locus of self-knowledge to its farthest reaches. Even admitting that parents may know better than we ourselves about our moral, intellectual, and aesthetic qualities; and even admitting that they may know better than even our closest friends what we really "feel and think deep down inside;" is it conceivable that any other being—however wise, however insightful—can know better than *we ourselves* what our inner thoughts and feelings are?

After asking: "Does your mother really know what you are like deep down inside?" "Does your father really know what you are like deep down inside?" and so on, we asked: "Now who knows best what kind of person you really are—your mother, your father, yourself, or your best friend?" As we would expect, older children are considerably more likely than younger ones to cite their "best friend" (16 percent to 4 percent). Let us, however, consider just those subjects who cited either a parent or the self as the locus of inner self-knowledge.

From the perspective of the above observations, it is somewhat astonishing to learn that the *younger children were as likely to attribute such knowledge to their mothers or fathers as to themselves.* As Table 10–5 shows, 52 percent of the younger children said a parent knew best what they were really like deep down inside, whereas 48 percent considered themselves the best judges on this matter. In part, this finding probably reflects a lesser awareness of their inner selves among young children, as shown in chapter 8, but in part it reflects a faith in adults bordering on the religious. Among the older subjects, these figures are reversed: 76 percent attribute prime knowledge to themselves, 24 percent to their mothers or fathers. The truth about the self is now to be found overwhelmingly within the self or a close friend; it is no longer vested in adults.

*TABLE 10-5.

*Locus of Inner Self-Knowledge, by Age*

|  | Age | | |
|---|---|---|---|
|  | 8-11 | 12-14 | 15+ |
| *Knows you best deep down* |  |  |  |
| Parents | 52% | 36% | 24% |
| Self | 48 | 64 | 76 |
| N = 100% | (489) | (383) | (266) |

*See footnote 7, chapter 3, p. 98.

RESPECT FOR AUTHORITY

Younger and older children thus differ radically in their views of where the ultimate truth about themselves is to be found. The young child is much more likely to feel that adults know better than even he himself how good, smart, or good-looking he is; that adults know better than his friends and siblings his innermost thoughts and feelings and that adults know even better than he himself what he feels and thinks deep down inside. The question is: is this remarkable respect of children for adult knowledge confined to the subject of the self or is it part of a broader set of attitudes toward adults that change over time? In order to examine this issue within the framework of more general attitudes toward grownups, we asked our subjects to agree or disagree with the following three statements:

1. Parents and teachers are always right.
2. The most important thing is to obey parents and teachers.
3. You should always do what parents and teachers tell you to.

These items were combined to form a "respect for adult authority" index. If there has ever been any question about Piaget's (1929: 378) claim that young children are awe-struck by authority—that they

**\*TABLE 10-6.**

*Respect for Adult Authority, by Age*

|  | Age | | |
|---|---|---|---|
|  | 8-11 | 12-14 | 15+ |
| *Respect for adult authority* |  |  |  |
| Low | 6% | 25% | 55% |
| Medium | 31 | 45 | 33 |
| High | 63 | 30 | 12 |
| N = 100% | (515) | (385) | (276) |

\*See footnote 7, chapter 3, p. 98.

view hierarchical authority as just, that they acknowledge the moral and scientific wisdom of adults—Table 10–6 may help lay such doubts to rest. It is rare indeed to find so powerful a relationship between a demographic and an attitudinal variable (gamma = .6516). Over three-fifths (63 percent) of the younger children, but less than one-eighth (12 percent) of the older ones, showed this awe of adult authority. Conversely, over one-half of the adolescents, but only one-sixteenth of the younger children, were relatively lacking in such respect.[1]

In order to test whether the relationship between age and the locus

**TABLE 10-7.**

*Zero-Order and First-Order Measures of Association Between Age and Locus of Self-Knowledge, Controlling on Respect for Adult Authority*

|  |  | Zero-order gamma | First-order partial gamma | % reduction | Zero-order r | First-order partial r | % reduction |
|---|---|---|---|---|---|---|---|
| *Relationship of age to . . .* | | | | | | | |
| 1) | Locus of exterior self-knowledge | \*.450 | .283 | 37% | \*.390 | .229 | 41% |
| 2) | Locus of inner self-knowledge | | | | | | |
| a) | Adults and Peers | | | | | | |
|  | Faith in Adults | \*.384 | .311 | 19% | \*.305 | .218 | 29% |
|  | Faith in Peers | \*.242 | .196 | 19% | \*.191 | .149 | 22% |
| b) | Adults versus self | \*.360 | .215 | 40% | \*.227 | .126 | 44% |

\*See footnote 7, chapter 3, p. 98.

of self-knowledge is in fact due to the young child's general respect for adult authority, Table 10–7 compares the original (zero-order) relationship of age to each of the four "locus of self-knowledge" measures with the first order relationships controlled on "respect for adult authority." [2] (In order to insure that the conclusions are not dependent on a particular measure of partial association, both partial correlations and partial gammas are included. These measures of association, it is evident, yield very similar results).

Table 10–7 indicates that respect for adult authority accounts in considerable part, but not completely, for the zero-order relationships between age and the locus of self-knowledge. The effect varies with the aspect of the self under consideration. About two-fifths of the relationship between age and the locus of exterior self-knowledge, and between age and interior self-knowledge, is explained by the striking respect for adult authority among the young. With regard to the relationship between age and faith in parents and faith in peers, about one-fifth of the relationship is apparently due to respect for adult authority. All these findings hold about equally for both black and white children considered separately.

## Discussion

In the early years, we have seen, the child is convinced that knowledge about both the exterior and interior self is vested in adults, whereas in the course of time the adolescent comes to believe that either a close friend or he himself is the ultimate source of truth on this subject. Three questions come to mind in reflecting on these findings: (1) Why are younger children so disposed to endow parents with such awesome knowledge about the self? (2) Why are older children so disposed to withdraw this respect and relocate the locus of self-knowledge within the self? (3) What is the significance of these facts for self-concept formation?

The reason the young child has such immense faith in adult knowledge of himself has probably best been described by Piaget. In the early years, Piaget (1948) contended, truth tends to be vested in sources exterior to the self. At this stage, it is the adult who is the repository of wisdom regarding the rules of the game, the origin of the earth, the principles of physics and mathematics, the problems of good and evil. If the child grants that the adult knows more than he does about these things, this might explain his willingness to grant to the adult greater knowledge of himself as well.

In addition, the young child's respect for adult insight is not entirely without foundation; the young child believes adults know his inner thoughts and feelings partly because they actually do. If the mother knows the child is hungry or sleepy without his saying so, it can only be because she can read his mind. When the young child says that adults know what he feels, thinks, and wants deep down inside, it is not because he has intentionally communicated these sentiments since, in fact, he is not yet disposed to share his thoughts with others. Adults understand him, then, not because he has bared his soul but because they *know*. The prototype of such a relationship is the child's relationship with God; there is actually no need to tell God anything nor is there any possibility of concealing even one's deepest and most secretive thoughts from Him because He is omniscient. Similarly, adults know our inner self because they are endowed with x-ray vision, with uncanny insight.

Third, in dealing with whether the parent or he himself knows what he is like deep down inside requires an assessment of how much he himself knows. The young child may feel that he knows little about what goes on "deep down inside" because, we have seen, he pays little attention to his psychological interior. He does not constantly reflect on his own thoughts, probe his own motives, examine his own feelings. Furthermore, it is noteworthy that he attributes to his parents an equal lack of interest in his inner states. Our subjects, for example, were asked to agree or disagree with the following statements: (1) "A kid told me: 'My mother cares more about how well I behave than how happy I am.' Do you feel like this?" (2) "Do you feel like this: 'My mother cares more about what I look like than how happy I am.'" The young children were far more likely than the adolescents to agree with these statements.

The adolescent's belief that he (or a friend) is the final authority on the subject of what he is like "deep down inside" is, of course, partly attributable to increased maturation, making the adolescent far more aware of adult fallibility. More to the point, however, is that the development of the ability to introspect, and the process of taking the role of the other, casts the entire concept of a psychological interior in a new perspective. As we saw in chapter 8, when the adolescent thinks of what he is like deep inside, he has in mind an inner world of thoughts, feelings, and wishes. But why are these inner states assumed to be imperfectly accessible to the parents? Because he knows that their (or anybody else's) inner states are imperfectly accessible to him. Just as he knows that ultimately he can never enter the mind of another with perfect accuracy, so, putting himself in the position of the other, he knows they can never enter his. Unlike the young child, the adolescent has achieved sufficient appreciation of interpersonal processes to understand that, ultimately, no one can ever know for sure what another person thinks and feels. Indeed, given the nature of parent-adolescent relations, the adolescent may be at pains to insure that this inner world *remains* inaccessible. Such concealment at once shows a keen awareness of a psychological interior on the part of the adolescent and a developed tendency to perceive himself from the adult's perspective. Thus, the process of taking the role of the other, already seen to account in considerable measure for how the individual thinks of himself and how he feels about himself, also helps to explain changes in the locus of self-knowledge.

This does not mean that the older child lacks respect for the adult but simply that his respect is qualified. In other words, even if older and younger children believe that adults understand them, the younger child may do so because of his awe of adult authority whereas the older child may do so out of sincere respect for the adult's insight, with full awareness of its strengths and limitations. Respect for adult insight into the self among different age groups may thus rest on different foundations—the first a childlike faith, the second a sincere respect founded on experience, evidence, and logic. But these are different types of faith. The physicist who believes that gas expands when heated does so with as much conviction as someone who believes that the earth was created in six days. The convictions may be equally strong, but the bases of the beliefs are different.

It is important to point out that the personality consequences of these facts may be lifelong and profound. Powerful as the adult locus of self-knowledge would be under any circumstances, it is further accentuated by the fact that, in the early years, the self-concept is relatively unstructured. This, then, may be one neglected reason for the importance of childhood for personality formation. Significant adults affect us for the following reason: At precisely the time of life when we are most unsure of what we are like, we are also at a stage where we stand most in awe of the moral and scientific wisdom of the adult. Hence, what adults think of us at this time of life (or, to be more exact, what we believe they think of us) is inordinately influential in determining what we think of ourselves; these self-concepts, in turn, govern our future—not totally, to be sure, but in large measure. Changes in the self-concept do take place, as we have seen in the previous two chapters, but they run squarely counter to the self-consistency motive, and some components may be remarkably resistant to change.

## Self-Concept Development: Summary

The previous three chapters have focused on changes in the nature of the self-concept or in interpersonal influences acting on it from the period from middle childhood (from about 8–9) through late adolescence (18–19). Only a small number of the varied aspects of the self-concept have been examined, but even these show sharp and fundamental changes as children grow older.

To recapitulate the findings of these three chapters, the Baltimore data have shown that when school pupils are presented with a series of general open-ended questions offering them the opportunity to conceptualize the self in any way they wish, their answers, broadly speaking, fall into any of seven categories: as a physical or material self (body, features, appearance); as a conglomerate of social statuses (sex, age, race, nationality, or other nouns); as a behaver or actor (crosses the street alone, listens to music, plays Greek dodge ball, cooks out on weekends); as someone with characteristic competence (good at arithmetic, can ride a bike, excels at fighting); as someone with typical interests and attitudes (likes adult television programs, usually takes a conservative position, likes grilled cheese); as a system of traits (honest, energetic, quiet), including interpersonal traits (likeable, easy to get along with, friendly); and as someone with certain inner thoughts and feelings (have a yen for Nancy, dream of living in Tahiti, resent my father).

*Younger Children's Self-concepts.* The self-concept of the younger child is particularly likely to consist of elements of a social exterior —relatively specific components which can be readily observed and require no probing or sophisticated synthesis. The younger child, spontaneously conceptualizing the self, tends to see it in terms of physical characteristics (short-tall, blonde-brunette, strong-weak), social identity elements (boy-girl, black-white, child-adult, American-foreigner), and relatively specific actions, abilities, or interests (rides a bike, plays hop scotch, messes his room, crosses the street alone, does poorly on long division). These are the things he is particularly proud or ashamed of, the ways in which he sees himself as similar to or different from others.

In addition, the self-concept at this stage of development is a relatively satisfied, stable, unreflective one. This quality of stability and acceptance, however, is based on a relative lack of interest in or concern with the subject-matter. Instead, attention is turned outward toward the interesting and important activities of life—play, school, television—not inward toward self-discovery. Not yet viewing himself from the perspective of others, the child has only a rudimentary propensity to view himself from the perspective of the "me," to see himself as an object. Hence, self-consciousness is generally low, self-concept stability high, and self-esteem satisfactory.

Third, the child's conclusions about what he is like rest very heavily upon the perceived judgments of external authority, particularly adult authority. Respect for peers—seen as small and incompetent— and even respect for his own judgment of what he is like does not match his faith in adults' views; knowledge of the self is regarded as absolute, and resides in those with superior wisdom. And these external authorities are assumed to share his own lack of interest in his inner states.

*Older Children's Self-concepts.* In early adolescence, the self-concept changes, and many of these changes persist into later adolescence. The earlier major components of the self-concept—physical characteristics, social identity elements, specific habits or interests— persist, but they are increasingly relegated to the background. Entering more centrally into the self-concept are, first, a psychological interior —an awareness of an inner world of thought, feeling, and experience. The self comes to be seen as a person with private thoughts and feelings, in general or toward other people. Second, the self is increasingly characterized in terms of abstract traits. This is true even among adolescents unfamiliar with, or unaccustomed to, trait terms. Third, the self is conceptualized as an interpersonal actor, and comes to be defined from the viewpoints of others.

Early adolescence, it appears, is a period of self-concept disturbance. Self-consciousness—especially an uncomfortable awareness of what others think about the self—rises sharply; the earlier period of unreflective self-acceptance vanishes. The self becomes more shaky, volatile, and evanescent. What were formerly unquestioned self-truths now become problematic self-hypotheses and the search for the truth about the self is on. Global self-esteem, as well as assessment of specific self-components, declines. In later adolescence, global self-esteem improves but, in general, self-concept disturbance persists. Whether the confident and unproblematic self-concept of the earlier years returns later in life—and, if so, how late—remains a matter for further research.

At the same time, the locus of self-knowledge shifts from without to within. The truth about the self, formerly the province of an all-knowing authority, now is vested in those in whom we have deliberately chosen to bare our inner selves—our best friends—or, particularly, ourselves. The recognition emerges that we alone have direct access to the invisible and intimate regions of the self. At the same time that respect for parental knowledge declines, the adolescent's confidence in his own expertise increases, causing a sharp shift in the locus of self-knowledge from without to within.

## NOTES

1. Our data, of course, do not represent tests of Piaget's more specific predictions, in part because our children are at least 8 years old whereas Piaget's studies also included younger children. The cognitive processes described certainly begin at an earlier age but, in our relatively mature sample, continue in an attenuated form.

2. The minor discrepancies between the zero-order gammas in Table 7 and those in Tables 1, 2, 3, and 5 are due to the fact that the Table 7 zero-order gammas are based on the identical cases as the partial gammas; that is, only those who had "respect for adult authority" scores. Those who failed to answer any "respect for adult authority" items are excluded from Table 7 but are included in the earlier tables.

# PART IV

# Beyond Self-Esteem

PART IV of this book consists of a single chapter. Its aim is to make explicit two ideas which have been left implicit at various points in the previous chapters, and to review and expand on them. The first is the importance of viewing the self as an active coping agent rather than as an inert, passive object, docilely obedient to the commands of external influences. The second is the need to recognize that the self-concept involves a great deal more than self-esteem, and to highlight the importance of examining other components, dimensions, or regions of the self. Data from the previous chapters will be drawn upon to illustrate these points and will serve as the springboards for the expanded discussion.

# 11

---

# Beyond Self-Esteem

---

**F**EW ACTIVITIES engage our lives so profoundly as the defense and enhancement of the self. The self-esteem motive intrudes on many of our daily activities, influencing what we say, how we act, what we attend to, how we direct our efforts, how we respond to stimuli. The individual is constantly on the alert, dodging, protecting, feinting, distorting, denying, forestalling, and coping with potential threats to his self-esteem.

The self-esteem motive is thus a constant force in our day-to-day lives. As Gordon Allport (1961: 155–6) said: "Every day we experience grave threats to our self-esteem: we feel inferior, guilty, insecure, unloved. Not only big things but little things put us in the wrong; we trip up in an examination, we make a social boner, we dress inappropriately for an occasion. The ego sweats. We suffer discomfort, perhaps anxiety, and we hasten to repair the narcissistic wound."

In the late forties, Gardner Murphy (1947) called to attention an important point, later reiterated by Hilgard (1949) and Allport (1961), that the various psychoanalytic defense mechanisms had as their most important single objective the protection of self-esteem. The mechanisms are familiar and were discussed briefly in chapter 2. Important though these mechanisms are, the single most powerful mechanism for self-protection and self-enhancement, we believe, is *selectivity* [1]—the motivated choice from among available options.

## Selectivity

In this chapter, we shall attempt to show the varied manifestations of selectivity with reference to the four principles invoked throughout this work: reflected appraisals; psychological centrality; self-attribution; and social comparison processes.

REFLECTED APPRAISALS

The looking-glass self version of the reflected appraisals principle holds that the individual's self-concept is importantly influenced by his perception of other's attitudes toward him. The question is: how do people exercise the selectivity mechanism to insure that the perceived self—the self seen through the eyes of others—is an attractive one? [2] Three mechanisms are suggested: selective interaction, selective imputation, and selective valuation and credibility.

*Selective Interaction.* One may confidently advance as a fundamental principle of social life that people, when given the choice, will tend to associate with those who think well of them and to avoid those who dislike them, thereby ensuring that the communications about themselves to which they are exposed are favorably biased.

The outstanding case in point is friendship. Friendship is the purest illustration of picking one's propaganda, for it is characteristic of a friend that not only do we like him but he likes us. To some extent, at least, it is probable that we like him because he likes us. Indeed, it is well-nigh impossible to be friends with someone who dislikes us, not only because we would have no taste for such a friendship but also because he would not allow the friendship to exist. The upshot of friendship selection is thus to expose people to implicit or explicit interpersonal communications which reflect well on themselves whereas they hear much less from people who dislike them.

People also confirm their self-concepts by selectively associating with those who see them as they see themselves. This is clearly shown in a study by Backman and Secord (1962). The subjects were 31 girls in a sorority house who were asked to describe themselves on 16 paired traits, such as warm-cold, bright-dull, mature-immature, dominant-submissive, and to rank the five adjectives most characteristic of the self. In addition, a "reflected self" check list was provided, asking each subject to indicate the five adjectives that she thought each other girl in the sorority would assign to her. The subject was also asked to rank everyone else in terms of how much she liked and interacted with the other girl. The data showed that the subject interacted most frequently with those other girls who, she believed, saw her as she saw herself. In addition, if the subject liked another person, she was more likely to believe the other saw her as she saw herself. The net result was that the person ended up having most involvement with those who saw her as she saw herself and least with those who saw her differently. This study is one illustration of the empirically demonstrated principle that people tend to like those who think well of them and to select friends or groups whose views of them correspond to their own (Dittes, 1959; Backman and Secord, 1959).

*Selective Imputation.* In emphasizing that the "looking-glass self" is "an *imputed* sentiment, the imagined effect of this reflection upon another mind," Cooley directed attention to the fact that the looking-glass self is a guess, a judgment of what goes on in the mind of the other. One would expect such guesses to be selective in the interest of self-esteem enhancement; given an opportunity to guess and a wish to protect self-esteem, it is likely that we will tend to ascribe to others more positive attitudes toward ourselves than they actually hold.

An interesting example of selective imputation appears in a study by Reeder, Donohue, and Biblarz (1960). As described in chapter 2, 54 enlisted military personnel were asked to rate themselves and other soldiers as a "leader" and as a "worker." In general, those rated high by others also rated themselves high and those rated low by others rated themselves low. But consider Table 11–1, which shows the relationship between the soldier's "objective group rank" (how he is actually rated as a leader by others) and "estimated objective group rank" (how he thinks others rate him as a leader). It can be seen that 24 out of 54 soldiers were mistaken with regard to the group's rating of them. These mistakes, it turns out, were far from random. Of these 24 "distorters," 21 believed that others ranked them *higher* than was actually the case whereas only 3 had the opposite impression. In fact, 15 out of 31 soldiers ranked "low" by others believed that others ranked them "medium" or "high." On the other hand, those who actually ranked high were most accurate in estimating the views of others toward themselves; only 2 out of 11 believed they were rated lower than they in fact were. We perceive other's views of us accurately if those attitudes are favorable but inaccurately if they are unfavorable.

**TABLE 11-1.**

*Estimated Objective Group Rank by Objective Group Rank ("Leadership" Rating)*

|  | Objective Group Rank | | | |
|---|---|---|---|---|
|  | High | Medium | Low | Total cases |
| *Estimated objective group rank* |  |  |  |  |
| High | 9 | 6 | 6 | 21 |
| Medium | 1 | 5 | 9 | 15 |
| Low | 1 | 1 | 16 | 18 |
| Total cases | 11 | 12 | 31 | 54 |

Source: L. G. Reeder, G. A. Donohue, and A. Biblarz. "Conceptions of self and others." *American Journal of Sociology* 66, 1950: 153-159, Table 8 (adapted). (Reprinted with permission of the University of Chicago Press. Copyright © 1960 by The University of Chicago.)

The same type of estimation applies to our beliefs regarding society's attitude toward our groups. As we observed in chapter 6, both the Baltimore data and other studies show that people believe that their groups (racial, religious, ethnic) are more highly regarded (liked, respected) by the society as a whole than is actually the case.

What holds true for social and religious groups appears to hold for occupational groups as well. Some interesting evidence appears in the well-known North and Hatt (1953) study of occupational prestige. Like racial and ethnic group status, there is general agreement on the ranking of the various groups in society; on the other hand, these estimates of prestige are subject to varying interpretation. Under these conditions, people may be motivated to misperceive the actual rank of *their own* occupational group. And that is what happens; there is a definite tendency for people to believe that their own or a similar occupation is generally more highly respected in the society as a whole than is actually the case.

This same tendency persists in our estimations of our *ego-extensions*.

In the Baltimore study, we not only asked the children how they ranked occupations in general, but also how they thought their fathers' occupations were ranked in the society. Although the rank order of occupations is surprisingly well-structured in the minds of even young children (Simmons and Rosenberg, 1971), there is one general exception to this principle—their fathers' occupations. These occupations are believed by children to command a higher regard in the society as a whole than they actually do. The status of races, religions, occupations, and fathers' occupations are, we see, "inflated."

There is thus evidence to indicate that people impute to others a higher regard for their traits, groups, or ego-extensions than these others actually hold—a misperception which fits comfortably with the wish to protect and enhance self-esteem. Although the individual sees himself through the eyes of others, what he tends to see is a more attractive picture than the one that is actually there.

*Selective Valuation and Credibility.* That significant others' attitudes toward us affect our own has been repeatedly demonstrated both in our own research and that of other investigators. But, as we stressed in chapter 3, not all others are equally significant to us. First, we *care* more about some people's opinions than others', and thus their opinions of us count more in shaping our own. Second, we *respect* certain people's opinions of us more than others', with like results. Hence it is not only the looking-glass self, but the *valuation* and *respect* of different others, that may affect the self-concept. But what makes some people more significant to us than others? Such significance, we believe, is selective and is chosen in the interest of protecting self-esteem and maintaining self-consistency.

Manis (1955) studied 101 college men living in a dormitory who were assigned to rooms on a random basis. The subjects described each member of their group, including themselves, on 24 bipolar rating scales. The subjects also made sociometric choices from among the members of their groups. The questionnaires were given twice, six weeks apart. One hypothesis of this study was that over time people would be more influenced by their friend's view than by the view of others. The results supported the hypothesis. The individual's self-attitudes became more similar to those of his friend than those of other people; this was particularly the case if his friend's views were favorable. Thus, our self-attitudes are more heavily influenced by those people whose judgments we value and trust; but we particularly value and trust those whose judgments of us are positive.

In chapter 3, it will be recalled, the Baltimore subjects were asked to indicate what they believed their parents, teachers, and best friends would say about them to others. The respondents were then asked: "You told me a moment ago what your parents would say about you. Would your parents be mostly right, somewhat right, or mostly wrong in what they would say about you?" Similarly, the respondents were asked whether their teachers and best friend would be mostly right, somewhat right, or mostly wrong. The results showed that, in every case, the child was more likely to believe the other was "mostly right" if the perceived self was favorable than if it was unfavorable. People are thus considerably more likely to value and respect the opinions of people who think well of them than of those who do not.

In sum, although the reflected appraisals principle correctly holds that we tend to see ourselves as we are seen by others, it is still possible to end up with self-attitudes considerably more favorable than those actually held by other people. An important reason is to be found in the mechanism of selectivity. First, to the extent that we can, we associate with those who like and respect us; adopting the attitudes of these particular others toward the self, the picture we internalize is likely to be a favorable one. Second, since taking the attitude of others toward the self is a matter of imputation, we tend to infer that others regard us more highly than they in fact do. Finally, there is the important question of whose views concern us and whose views we trust. Since interpersonal significance is selective, we tend to accord significance to those who, in our view, hold positive attitudes toward us while withholding it from those who dislike us. By means of these three selectivity mechanisms, the individual, following the principle of reflected appraisals, is heavily influenced by what the people with whom he interacts think of him, and yet ends up with a level of self-regard considerably higher than the average level of respect in which he is actually held by others.

PSYCHOLOGICAL CENTRALITY

Not all self-components, we have stressed repeatedly, matter equally to the individual; some are central to his feeling of worth, others matters of indifference. In light of this fact, how does the individual proceed to protect his self-esteem? One way to do so, it is apparent, is to draw to the center of his value system those qualities at which he feels he excels and to relegate to the periphery his areas of incompetence. The upshot is that the good things are the ones that matter and the bad things the ones that do not. In illustrating this point, it is convenient to distinguish (1) selectivity of self-values, and (2) selective introjection.

*Selectivity of Values.* As stressed in chapter 1, some components of the self are central, whereas others are matters of relative indifference. A scientist may be very much concerned about his scientific ability but not about his typing skill; for a secretary, the reverse may hold. Hence, aside from the objective facts used to reach conclusions about one's scientific or typing ability, there is the further issue of how *important* these components are to the individual. Even if there were an objective reality which could establish how good a scientist one is, there is no objective reality which can prove to the individual that he should *care* whether he is good at science—whether this is a major self-value.

Consider the individual's traits or dispositions. We have seen in chapter 2 that the individual's assessment of his *specific* self-concept components and *global* feeling of self-worth tend to be related. But the strength of this relationship depends on the *value* attached to the specific quality. Among New York State adolescents, if the youngster cared greatly about the quality, then the relationship between the specific assessment and global self-esteem was much stronger than if he did not care.

The same finding, it will be recalled from chapter 5, appeared among Chicago adults. Among those who cared greatly about money, the relationship between objective income and global self-esteem was relatively strong, but among those to whom money mattered less, the relationship was much weaker. The same held true regarding the association of occupational prestige to global self-esteem.

But this finding raises the obvious question: what about the self does the individual elect to value? One hypothesis is that he will be disposed to *value those things at which he considers himself good and to devalue those qualities at which he considers himself poor.* As one illustration, consider the quality "good at working with your hands." Table 11–2 shows that, among those who felt they possessed this skill, 68 percent valued the quality highly, whereas among those who believed they lacked this quality, only 6 percent attached that much importance to it. Examining 16 qualities in New York State, we found that, *without exception,* those who considered themselves good in terms of these qualities were more likely to value them

**\*TABLE 11-2.**

*Self-Estimate and Self-Value for Quality*
*"Good at Working with Hands"*

|  | Self-Estimate: Actually Consider Self "Good at Working with Hands" | | |
|---|---|---|---|
|  | Very | Fairly | Little or Not at All |
| *Self-Value: Care About Working with Hands* |  |  |  |
| Care a great deal | 68% | 27% | 6% |
| Care somewhat or little | 32 | 73 | 94 |
| N = 100% | (224) | (392) | (533) |

\*See footnote 7, chapter 3, p. 98.

than those who considered themselves poor. Self-values, we see, tend to be selected in such a way as to enable the individual to maintain a congenial self-picture.

Of course, the arrangement works both ways: The individual will certainly seek to excel at those things about which he cares. Someone who cares greatly about his musical talent may spend endless hours practicing and studying in an effort to excel at that which he values. But the reverse also applies. Consequently, we not only seek to excel in those areas on which we have staked ourselves but we tend to stake ourselves on those areas in which we excel. We may devote as much ingenuity, effort, and energy in searching for virtue as in cultivating it.

The selective valuation of traits also applies to physical characteristics. Although most black children in the Baltimore study said that all skin colors were equally nice, indirect evidence showed that light skin color was considered more aesthetic (see chapter 6). What does the darker-skinned child do under these circumstances? One way he reacts to his predicament is to decide that good looks simply matter less to him (that perhaps intelligence or morality or character are the things that really count.) The data show that 38 percent of those saying they are "very dark," but 24 percent of those describing themselves as "very light," said that they cared little about whether they were good-looking. Unlike most traits, which are, at least in principle, changeable, there is nothing one can do to change skin color. What one can do, however, is to decide that looks don't matter; and this is precisely what the darker-skinned child is more disposed to do.

In our usual way of thinking, we assume that our assets are sources of pride, our defects sources of shame. To a considerable extent, however, our assets are indeed sources of pride but our defects, not sources of shame but of indifference. Thus, in totalling up our sum of virtue, we place the greatest weight on our assets, least on our debits. This accounting system both protects and enhances self-esteem. Furthermore, as Tamotsu Shibutani points out (1961: 436), "There are an amazing variety of attributes of which people are proud or ashamed: their speaking voices, the straightness of their teeth, their ancestry, their muscular strength, their ability to fight, the number of books they have read, the number of prominent people they know, their honesty, . . . their ability to manipulate other people, the accessories on their automobiles, or their acquaintance with exotic foods." These assets are easy to find.

*Selective Introjection.* The boundaries of the self, we have observed, tend to be nebulous and changeable; in the course of time new objects or individuals physically external to the self are experienced as a part of the self whereas others are sloughed off or relegated to the fringes. Furthermore, if the external component is indeed incorporated into the self and constitutes an ego-extension, then the individual's positive or negative attitude toward that object will have a stronger self-esteem effect than if it is not. For example, in the discussion of racial and religious group identification in chapter 7, it was shown that if the individual strongly introjected his race or religion—considered it an integral and essential part of the self— then his degree of pride in that group had a stronger impact on his global self-esteem than if the group was not introjected.

Which groups or ego-extensions, then, will the individual be disposed to introject? As the data in chapter 7 clearly indicated, he will make more central to his sense of self those ego-extensions of which he is particularly proud, and relegate to the periphery those of which he is not. For example, 82 percent of those black children who felt great pride in their race would take it personally if their race were insulted, but only 31 percent of those who felt little pride would consider an attack on the race as an attack on the self. For religion, the corresponding figures were 83 percent and 44 percent. To the extent that the individual is able, he will elect to incorporate into his sense of self those external objects which are, in his view, particularly admirable but extrude those which are not.

In sum, the principle of psychological centrality holds that those traits, physical characteristics, social identity elements or ego-extensions which have the greatest importance for us will have a stronger impact on our global self-esteem than those to which we are indifferent. But what ranks highest in our self-value system or is permitted to become an integral part of the self is subject to the mechanism of selectivity. The structure of the self-concept is thus in part the product of selective influences operating to protect self-esteem.

## SELF-ATTRIBUTION

"Attribution theory" (Kelley, 1967) is largely concerned with examining the cognitive processes of people when they attribute "causes, dispositions, or inherent properties" to entities on the basis of observation of their characteristics and behavior. Attribution, it is evident, involves not only observation but interpretation as well; and this is as characteristic of *self*-attribution as of attribution to any other entity. Such interpretation, we believe, is both motivated and selective. The conclusions we draw about ourselves by observing our behavior are influenced by at least four kinds of selectivity: (1) selective causal attribution; (2) selective interpretation of facts; (3) selective attention to facts; and (4) terminological selectivity.

*Selective Causal Attribution.* In examining the question of cause, attribution theorists have stressed the distinction between internal and external attribution. If I like something, is this because the thing is good (object desirability) or because of my disposition to enjoy such things (personal desire)? If I lose a tennis match, is this because I played poorly or because my opponent played well? If I fail a test, does this mean I am stupid or that the test was unfair?

The research evidence is consistent in showing that such attribution is highly selective. An early study by Johnson, Feigenbaum, and Weiby (1964) experimentally manipulated "success" or "failure" on a task, and then asked subjects why they had failed or succeeded. Those told they had done poorly explained their performance in terms of external conditions, whereas those told they had done well explained it in terms of internal characteristics. The self-esteem of the successful is thus enhanced, the self-esteem of the unsuccessful protected.

This pattern of selective causal attribution is a general one (although exceptions appear). Discussing four studies which "have examined the causal attributions of individuals who received false feedback that they had succeeded or failed at a task," Wortman *et al.* (1973: 372) concluded that "these studies have generally found that individuals attribute causality to themselves when they succeed while attributing causality for their failures to facts in the environment or situation."

In the experiment by Wortman *et al.* (1973), the subjects were given a test of "social perceptiveness" and told that they had done either very well or very poorly. How did those told they had done "poorly" account for their poor performance?

The first device was to attribute their poor performance to impersonal and uncontrollable forces, such as luck; by implication, the successful subjects attributed their good performance to personal merit.

A second tendency was to explain the poor performance by reference to the objective difficulty of the task. True, they had failed at the task—the experimenter, who knew the "truth," had told them that—but it was not because they were stupid but because (a) the questions were hard; (b) they had been given insufficient information to reach correct conclusions; or (c) the information given them was confusing.

The third stratagem was to attribute failure to lack of motivation rather than lack of ability. In comparison with the successful subjects, those who failed were more likely to deny that they had *tried* to do well. Failure thus did not reflect on inherent worth.

The fourth stratagem in the face of failure was to deny the importance of success. The failures were more likely to hold that how well they did in the "social perceptiveness test" was trivial or unimportant.

*Selective Interpretation.* Assume, however, that the individual does attribute his behavior or its outcomes to internal dispositions. What makes this process less direct than the self-perception theorist would have us believe is that, since the same dispositions may have diverse behavioral manifestations and different dispositions may be reflected in the same behavior, there is very considerable variation in the interpretation of observed facts about the self.

As an indication of the diversity of possible interpretations, consider one of the few relatively clear and unequivocal facts about the school pupil—his grade average. Grades are an objective measure of performance. But what disposition underlies the performance? This is largely up to the individual to decide. Consider a number of characteristics that reflect intellectual qualities: good student in school;

intelligent, a person with a good mind; clear-thinking and clever; imaginative and original; knowing quite a bit about many different things; a person with good sense and sound judgment; a logical, reasonable type of person.

The New York State data show striking differences in the degree of association between grade average and self-estimations on these qualities. The association of objective grades to the belief that one is a "good student in school" is .52, but the association of grades to the conviction that one is "clear-thinking and clever" is only .16, and the association of grades to the belief that one is "imaginative and original" is only .08. In other words, most people agree that grades are a good indication of whether or not they are good students, but they are by no means convinced that grades tell much about whether they are "clear-thinking and clever" or "imaginative and original." There are many facets of intelligence, and there is nothing in the objective situation to compel the student to interpret his poor grades as relevant to these aspects of intelligence.

Probably the most common, and certainly the most familiar, example of selective interpretation is the Freudian mechanism of rationalization—the selective search for reasons, justifications, excuses, and accounts of one's behavior or its consequences which cast the actions in a favorable light. Such rationalizations almost always have at least a grain of truth, and the individual selectively focuses on that aspect of the action or its outcomes which permits him to attribute favorable underlying dispositions to the self.

*Selective Attention to Facts.* Although we draw conclusions about our dispositions by observing our behavior (or related evidence about the self), it is equally true that which behavior we elect to judge and which to ignore is not random. With regard to any disposition, the number of indicators which reflect it are almost infinite. Intelligence, for example, may be reflected in test performance, a witty remark, solving crossword puzzles, making rational purchases, betting on the winning horse, understanding T. S. Eliot. But it is not necessary to do all these things to be deemed intelligent. The individual thus has some choice in the selection of indicators relevant to the disposition. Given this choice, it is likely that he will search for those items of evidence leading to congenial conclusions about the self, attempting at the same time to overlook or ignore those items of evidence which do not.

Self-esteem can thus be protected and enhanced not only by *ex-*

*plaining* facts and by *interpreting* facts but also by *attending* to facts about the self in the first place. This finding appears repeatedly in occupational studies. Simpson and Simpson (1959), in their study of psychiatric attendants, start with the thesis "that people in low status occupations may seize upon some aspect of the work which is highly valued throughout the society or in the work sub-culture and build a self-image around it." Having chosen an occupation perhaps for extrinsic rewards (such as money and security), people subsequently seek other values satisfied by the occupation. "Attendants tend to minimize the less glamorous features of their work and focus upon the most highly valued element in the sub-culture: care of the patient." Thus, the psychiatric attendants stressed their contribution to patient welfare, taking pride in their alleviation of human suffering. According to the authors, this was not why they had taken the job in the first place (money and security were the prime incentives), and this was not what they actually spent most of their time doing (such as cleaning and housekeeping). "The data thus support the hypothesis that people in a low status occupation can develop and maintain a favorable occupational self-image by focusing on some highly valued aspect of the work situation" (1959: 392).

Other occupational groups are equally selective. Seidman's (1962) study of telephone workers indicates that the telephone operator focuses on those aspects of the job which are superior to manual work—the cleanliness of the job, better manners, superior dress. "It's not like manual labor, it's more like office work." "It's the same as any business office. In fact, I think they [telephone operators] should be called communication secretaries because they do a great deal of work for business firms."

We thus advance as a general proposition the view that whenever an individual comes upon an object, either himself or an ego-extension, he will tend to scan that object with an eye to searching out its strengths, merits, and virtues, and will then amplify the importance of those components. This tendency is made possible by the fact that any object consists of a large number of aspects, elements, and dimensions or can be viewed from a variety of viewpoints. People show a striking proclivity to give particular attention to those facts, characteristics, or consequences which are admirable and to be less attentive to those not as good.

*Terminological Selectivity.* As stressed in chapter 1, the self-concept is a distinctively human creation and is constructed largely from the materials of natural language, that is, through the use of terms. Such terms, however, are usually biased and inexact reflections of the underlying disposition concepts they are intended to represent. Traits —smart, brave, boring, charming—are pertinent examples of such abstract concepts represented by terms. But the language of traits is not objective and value-neutral; on the contrary, it tends to be pregnant with evaluative connotations. Most trait terms describe not only action tendencies but also expressions of approval or disapproval.

Since this is the case, it is understandable, if people are motivated to protect and enhance the self-concept, that they would be disposed to apply to their behavior those adjectival terms that cast it in a flattering light. Horney offers a number of examples of this propensity in the neurotic (1950: 94): "His capacity for this unconscious reversal of values is perfectly amazing. Thus inconsistency turns into unlimited freedom, blind rebellion against an existing code of morals into being above common prejudice, a taboo on doing anything for oneself into saintly unselfishness, a need to appease into sheer goodness, dependency into love, exploiting others into astuteness. A capacity to assert egocentric claims appears as strength, vindictiveness as justice, frustrating techniques as a most intelligent weapon, aversion to work as 'successfully resisting the deadly habit of work,' and so on."

The same terminological selectivity applies to social identity elements. The practice of relabeling goes on incessantly. As we pointed out in chapter 1, one method of relabeling is to absorb the abjured category into a respected one (retardates become one of a class of "exceptional children"), the other to cast the socially abjured category into the morally neutral category of scientific terminology ("nuts" or "loonies" become "mental patients").

With regard to the universal practice of terminological selectivity, two points should be stressed. The first—and this cannot be too strongly emphasized—is that this practice does not necessarily represent a distortion of reality. The laudable term describes the objective reality with exactly as much (or as little) accuracy as the deprecatory one. The lunatic and the mental patient, the garbage collector and sanitary engineer, are one and the same; the term "grind" is no more (or less) accurate than "hard-working student," the term "jock" offers no more accurate description of reality than "athletically inclined." The appeal of this mechanism is apparent. It is a good deal easier—and may be equally ego-enhancing—to change the term as to change the person.[3]

The second point is that the content of the self-concept is to an important extent constituted from the terms of natural language. People do conceptualize themselves to an appreciable extent in these terms. This is not to say that they do so completely. Physical images or images of the self in action are also important self-concept components. Furthermore, the individual can view himself in terms of the abstract underlying disposition concept without reference to a particular term. But to a considerable extent, these disposition concepts are crystallized in the self-concept in the form of natural language, and the terms are characteristically selected with an eye to self-protection and self-enhancement.

In sum, there is nothing in the discussion above to challenge the self-attributionist's assertion that the individual's self-concept is to an important extent formed by observing his behavior or other visible facts about the self. But the entire process of moving from the observable act or fact to the self-concept is rife with selectivity. True, we do observe our behavior but whether the *cause* of the behavior is internal or external (causal attribution); what the behavior *means* about us (interpretation of facts); how much *attention* we pay to the particular behavior (attention to facts); and what terms we choose to describe the underlying disposition (terminological choice)—all these are governed in considerable part by selective processes working to protect self-esteem.

SOCIAL COMPARISON PROCESSES

In his discussion of the social evaluation principle, Pettigrew (1967) pointed out that the individual views himself with reference to at least two criteria: (1) in relation to other people—Festinger's "social comparison" principle, and (2) in relation to certain defined standards—Thibaut and Kelley's "comparison level." The individual's comparison of the self in these terms has obvious self-esteem implications; the question is whether or not selective processes influence such comparisons.

*Selective Social Comparison Processes.* Since the individual's view of himself as better or worse, superior or inferior, higher or lower than other people is an important basis for self-esteem, one question is whether these comparative judgments are accurate. The data suggest that people's judgments of themselves as above or below average is highly selective.

In his study of a group of business managers, French noted: "Ninety-two members of management were asked to evaluate their performance on a percentile scale with other men in the same unit doing the same job. . . . Only two of the ninety-two rated their performance as below the fiftieth percentile" (1968: 144). This remarkable statistical feat reappears constantly in research. Consider Heiss and Owen's (1972) study of black and white adults. Subjects were asked the following: "Now I would like you to rate yourself as above average, about average, or below average on some things that you do and some things that you are. First, would you say that as a son (daughter) you were above average, about average, or below average?" The question was repeated for nine other traits or social identity elements: as a parent; as a spouse; as a conversationalist; in attractiveness; as an athlete; in willingness to work hard; in trustworthiness; in intelligence; and in mechanical ability. It turned out that on six of the 10 items only about 2 percent of the subjects rated themselves below average, and on two others about 10 percent rated themselves this low; almost everyone is, in his own eyes, average or above average.

Apparently the same holds true for social identity elements, as clearly evidenced in the "group favorability bias" phenomenon discussed in chapter 6. Brewer and Campbell's (1976) study of East African groups showed each group considering itself best or one of the best, a finding confirmed by Harding *et al.* (1969) in their general review of the American literature. Even in terms of selected stereotypes, the Laurence (1970) study showed a substantial proportion of whites who considered whites superior to blacks but almost no blacks who thought blacks inferior to whites. According to their members, then, their groups are either as good as or better than most other groups; virtually no group, in its own view, is below average.

*Selective Comparison Levels.* People judge themselves not only in comparison to other people but also in comparison to certain standards. These may be the standards of the individual's own past performance, of some referent other (say, father or mother), of a reference group (for example, the adolescent peer group), or even the standards of James's "ideal judge." A man may aim to be a dominant figure in the world of business or politics or to be a competent plumber or carpenter. This principle is even more true of nonoccupational goals. One person may aspire to be the ultimate in sweetness, goodness, and kindness whereas another is content just to be a decent person. There is thus a great choice available in the setting of standards of performance in the great sweep of areas pertaining to the self.

Given these options, what standards do people select for themselves? Gordon Allport (1943: 470–471) has summarized the level of aspiration studies in a sentence. He commented: "Unless I am mistaken, every investigation has directly or indirectly confirmed Hoppe's initial claim that the subject behaves in such a manner as to maintain his self-esteem at the highest possible level."

## Discussion

In sum, if there is any general inference to be drawn from this discussion of selectivity, it is that the human animal is not a sponge blotting up the social forces impinging on him and spraying them forth as psychological and behavioral consequences. Rather, he is constantly in the process of reacting to, and coping with, the social events that constitute his experience in the service of his own system of motives. Interpersonal selectivity, selective imputation, selective significance, selectivity of values, selective introjection, selective causal attributions, selective interpretation, selective attention to facts, terminological selectivity, selective social comparisons, and selective comparison levels are only *some* of the selectivity devices people employ in the interest of their self-concept motives. People do not simply accept what exists but *react* to it, often in astonishingly fertile and ingenious ways. What they all add up to is a series of mental mechanisms attempting to interpret reality in such a way that both we and other people see ourselves as we would wish to be seen.

In addition, confronted with any threat, the individual need not confine his defense to a single selectivity mechanism but may employ an entire battery, simultaneously or sequentially. Matthew Erdelyi (1974) suggests that perceptual defense may be viewed as an information-processing system aligned in sequential stages, and that perceptual defense may appear at any stage in the sequence. The threatening information is not seen; if seen, it does not register; if registered, it is misinterpreted; if correctly interpreted, it is forgotten; and so on. It is as though the raw material of external reality had to pass through a series of ever finer sieves which filter out the impurities—those facts offensive to the self. Insofar as this filtering system is effective, what eventually remains within our phenomenal fields is material either consistent with, or enhancing to, our self-concepts.[4]

Furthermore, people do not simply sit back and attempt to cope with the forces that confront them, but they sally forth in an active effort to protect and enhance self-esteem. A person does not simply avoid the company of those who despise him but actively seeks out the company of those who respect him. A student does not simply avoid courses at which he expects to fail but also searches for courses at which he expects to excel. In short, people do not engage only in an active defense of, but also in an aggressive search for, self-esteem.

*Where Selectivity Fails.* Given the effective operation of a multitude of selectivity mechanisms, it might seem surprising that *anyone* should have low self-esteem. Regrettably, people do. There can be no question that many people have mild or moderate doubts about themselves, others serious doubts, and still others grave doubts. Such evidence does not mean that the principle of selectivity is wrong, but that there are certain realms of human experience in which it is not free to operate without limit.

The most obvious limitation, of course, is that there are certain objective facts about the self which cannot be evaded. Thus, it is clearly the case that students with low grades *are* less likely to consider themselves good students than those with high averages. Similarly, there is no way for a short man to think of himself as tall, a poor man as rich, or a lower-class person as an upper-class person so long as he is in contact with reality.

Second, although we characteristically elect to interact with those who think well of us, in many cases we are deprived of any choice in the matter. We may be relatively free to choose whom we wish as friends, but the same is not true of parents, teachers, or children, and far less true of neighbors, employers or employees, colleagues, or clients. Thus, if our parents scold us, our teachers greet our remarks with sarcasm, our fellow workers laugh at us, or our employer berates us, we are largely deprived of the option of avoiding their company or their criticism.

Third, we are not completely free to choose our self-values. Many self-values are acquired long before the opportunity to test them adequately is at hand. Thus, a child from a musical family may learn early to value musical skill, and it becomes apparent only later that he has insufficient talent. Self-values, particularly if established early and reinforced by significant others, may be quite difficult to change even if, at a later time, it is in the individual's interest to do so.

Another point is that in many cases *traits are to goals as means are to ends*. We cannot conveniently abandon the importance of a particular quality without at the same time abandoning the goal for which this quality is required. In the case of the doctor's son who wants to follow in his father's footsteps, the young man will consider it very important to possess the qualities necessary for academic success. It would be absurd to assume that he can dismiss this value and raise to priority the value of being a good dancer, a skill at which he excels. So long as he maintains the goal of becoming a doctor, he must be concerned with academic competence; he cannot waltz his way into medical school.

Finally, people cannot completely evade the values stemming from social role definitions and social group norms. At an early age the child learns what is right or wrong, important or unimportant, for him, and these ideas are internalized in his value system. In addition, as he grows up, he finds himself judged by these criteria, so that if he desires the approval of his group, he must seek to excel in terms of *their* values, not his own. True, he may momentarily enhance his self-esteem by abandoning their values, but this act is likely to call down on him the disapproval of his significant others, which would probably counterbalance any rise in self-esteem. Self-values are malleable but they are not completely subservient to the needs of psychological comfort.

In sum, although the chief weapon in the struggle for self-esteem and self-consistency is the mechanism of selectivity, there are many life situations in which selectivity fails, in which the use of this mechanism is impossible or leads to still more damaging consequences. But if people are able to employ the selectivity mechanisms in the service of their self-motives, they will.

## Beyond Self-Esteem

Future progress in self-concept research will not only require us to take account of the individual as a creative, coping organism, but will also require us to broaden our vision and to attempt to go beyond self-esteem.

This is not to imply that global self-esteem is unimportant; certainly, the significance of self-esteem is beyond dispute. This conclusion is evident not only in the discussion of the various selectivity mechanisms—only a very powerful motive could enlist this much human effort and ingenuity in its service—but also in the strong correlations between self-esteem and diverse manifestations of psychological disturbance described in chapter 2. If one wished to investigate a *single* aspect of the self-concept, then, global self-esteem would probably be the most important. Yet if we adopt the view proposed at the outset of this work that the self-concept is the totality of the individual's thoughts and feelings with reference to himself as an object, it is evident that the great bulk of it is, from the viewpoint of systematic research, largely untouched. Our aim is to call attention to some of these neglected areas.

NEGLECTED AREAS

*Content.* It is surprising how many investigators literally pay no attention to their own data. Subjects are asked to describe themselves in terms of a number of qualities or characteristics, but the resulting data are then *combined* (by factor analysis, the evaluative dimension of the semantic differential, self-ideal discrepancy score, or any of a number of other arithmetic or mathematical procedures) in search of an underlying common dimension of global self-esteem. But, as we stressed in chapter 1, the specific and the global measures are both important but are neither identical nor interchangeable. It is certainly relevant to know about the individual's *general* feeling of worth, but it is also important to know what he thinks of such *specific* qualities as his intelligence, athletic skills, kindliness, or morality. Furthermore, it seems apparent—and research confirms the expectation—that specific self-attitudes will usually more effectively predict specific behavior than will global self-regard. This obvious point is overlooked with surprising frequency. It is common, for example, to treat the academic self-concept and global self-esteem as interchangeable. Educators and others concerned with the poor academic performance of disadvantaged groups have frequently attributed these behavioral outcomes to self-concept problems and have called for actions designed to improve the global self-esteem of these children. Yet it should be evident that, insofar as self-concept factors are implicated in these particular results, they stem much more from negative intellectual or academic self-concepts than from low *global* self-esteem. For example, Bachman

(1970: 242) shows that the correlation of school marks to a "self-concept of school ability" measure is .4817 whereas the correlation of marks to global self-esteem is .2334. It is apparent that, insofar as the self-concept influences school marks or academic achievement, it does so less by virtue of a general feeling of self-regard than by the specific belief that one is intelligent or academically talented.

Conversely, the evidence suggests that, in general, sociologists and psychologists have been more successful in predicting specific self-concept components than in predicting global self-esteem. One broad conclusion suggested by Ruth Wylie's (1978) exhaustive survey of the substantive data is that clear and consistent relationships of sociodemographic variables and global self-esteem are not commonly found. On the other hand, there is evidence to indicate that contrasting groups (the old and young, rich and poor, boys and girls) do see themselves differing with regard to certain specific components. Treated thoughtfully, such differences could be of considerable interest and significance.

Particularly neglected self-concept components are "social types." Here we have in mind not the heuristic types of the social scientist but the empirical types recognized within groups. Characteristically, these types are syndromes, frequently combining normative practices, style of life components, values, interests, personality characteristics, and demographic characteristics. Though the typological term may be based on some central element, it connotes an idea broader than this particular component. Little is known about how these types are internalized in the self-concepts of members of different groups, or about their varied behavioral consequences.

There are, however, two serious dangers involved in research on specific self-concept components. The more serious danger is that of triviality. We are scarcely likely to experience the great thrill of scientific discovery at finding that top students are more likely than others to have positive academic self-concepts, or that those over 50 are more likely to consider themselves "middle-aged" than those under 25. Though a great deal of variance may be explained, such findings are not worth finding. But it does seem worth examining the degree to which interpersonal attraction can be understood with reference to an individual's self-picture as a "hail-fellow-well-met," or as someone "loaded with charm;" or the degree to which community participation is related to one's self-concept as an "activist" or "solid citizen." There is no lack of interesting and significant hypotheses that can be generated.

The second danger lies in attributing to the specific self-concept component, as a causal influence, an effect which is due to the *actual* self. For example, people who view themselves as neurotic are likely to experience interpersonal difficulty. But it may be that the interpersonal difficulty is attributable not to the fact that they *view* themselves as neurotic but to the fact that they actually *are* neurotic. The effect of the self-concept would only be suggested if, *at equal objective levels of neuroticism*, the self-concept and interpersonal difficulty were related.

This distinction is critical. Whether someone will try out for the team or apply to college depends on his self-concept; whether he will make the team or succeed in college depends on both his self-concept and his actual self. In conducting research on the self-concept, it is important either to focus on behavioral outcomes in which the self-concept rather than the actual self is clearly implicated (such as applying to college), or to examine relationships controlling on the actual self (for example, the relationship between academic self-concept and school performance, *controlling on IQ or achievement potential*). Brookover, Thomas and Paterson (1964), for example, in a study of 1,050 seventh-grade students, found that the zero-order correlations of academic self-concept and grade-point average was .57 for both males and females; controlled on IQ, it still remained .42 and .39, respectively. Thus, if those who think they are good students do well in school, it is not solely because they actually are more intelligent. This finding strengthens the likelihood that the self-concept plays an independent role (although the direction of causal influence is not unequivocal).

*Structure.* Even greater neglect has been accorded self-concept structure. Although writers occasionally speak of the structure of the self-concept, few appear to have taken the idea of structure seriously. Yet one can no more understand the self-concept by studying its traits, physical characteristics, and social identity elements in isolation than one can understand a watch by studying the gears, springs, and cogs that constitute it; it is the arrangement of the parts that constitutes the entity.

One area of self-concept structure particularly in need of attention is the hierarchical ordering of various elements. Assume we learn that someone considers himself a mediocre tennis player. In the absence of knowledge of the degree to which the self is invested in this characteristic, we can have little idea what the emotional and behavioral responses to this self-assessment will be. Can we predict that the individual will spend a good deal of time taking lessons, practicing his service, improving his ground strokes? Can we infer that he suffers feelings of chagrin and self-rage when he double faults? Obviously not, if we have no idea whether he cares an iota about tennis skill. But how many investigators ask their respondents how intelligent, good-looking, likeable, moral, or neurotic they are without ever bothering to determine how much the respondent cares about these characteristics, that is, where these characteristics rank in his hierarchy of values? If the hierarchical ordering of traits is neglected, the hierarchical ordering of social identity elements is even more so. One person takes great pride in his social class position, a second in his ethnic background, a third in his race, a fourth in his religious affiliation, and so on; the concept of "identity prominence" (McCall and Simmons, 1966) needs further research.

In the course of this work, we have seen several examples of how the contribution of each particular component to the totality depends on the importance, centrality, or priority of the particular component to the total feeling of worth. In chapter 2, we noted that when adolescents were asked to rate themselves on 16 traits and were also asked how much they cared about these qualities, in almost every case the relationship between the component self-rating and global self-esteem was stronger if the youngster cared greatly about that quality than if he did not. Similarly, in chapter 5 we saw that, among the Chicago adults, the size of the relationship between actual income and global self-esteem depended on the centrality of money in the individual's value system; and the relationship of occupation to self-esteem hinged on his concern with social prestige. Finally, in chapter 7, evidence was adduced to suggest that the relationship between pride in one's race or religion and global self-esteem depended on the degree to which these groups were introjected. Future work must certainly give consideration to the ordering, arrangement, and interrelationships of the various self-concept components, as well as to the relationship of each to the whole.

*Dimensions.* Most of the self-concept dimensions—stability, clarity, consistency, certitude, salience, self-confidence—have suffered similar neglect. The dimensions of direction and intensity, that is, self-esteem, have received the lion's share of attention; stability the rabbit's share, and the remainder almost no share.

Global self-concept salience—self-consciousness—is a conspicuous case in point. By salience we mean the degree to which the self as an object is prominent in one's mind, is at the forefront of attention. Self-consciousness, as each of us knows from his own experience, is an important dimension of the self-concept. It is evident in diverse behavior—nervousness at a public presentation, blushing, stammering, hand-trembling, embarrassment—but also in the presentation of a seemingly "false" presenting self, that is, one that appears to be uncongenial, lacking in spontaneity, artificial, or conspicuously intended for effect. Self-consciousness is also related to other indicators of emotional disturbance, especially to psychophysiological indicators of anxiety. Self-consciousness, then, is a dimension that is recognized in experience, implicated in behavior, and related to emotional disturbance. Furthermore, developmental and social structural factors influence it strongly. As we indicated in chapter 9, it is at its height in early adolescence, rising sharply from the unselfconscious levels of the earlier years, and leveling off in later adolescence. In addition, at each age level it is higher among girls than among boys. An adequate measure of this dimension still awaits us but it is an incontrovertibly important aspect of the self-concept in need of further research.

Although some attention has been given to self-concept stability in the literature, its actual importance has not been appreciated. People want not only positive self-attitudes but stable ones as well. The New York State study showed clear and definite associations between self-concept instability and certain signs of emotional disturbance, such as depression and psychophysiological indicators of anxiety.

Furthermore, we have seen that developmental, contextual, and social structural factors may play a role in influencing self-concept instability. First, chapter 9 shows that this instability is related to age. It is in early adolescence (ages 12–14), when the actual self is undergoing great physical change, when social experiences are altered as a consequence of moving from the elementary to the junior high school, and when a new and mature stage of social intimacy is achieved, that the self-concept appears to become more unstable. Second, this instability is related to the youngster's sex, with certain physiological changes and sex role definitions contributing to greater self-concept instability among girls (F. Rosenberg and Simmons, 1975). Third, our examination of social contexts in chapter 4 offered some indication that, self-esteem aside, contextual dissonance was associated with greater self-concept instability among youngsters of certain ages. Thus, self-concept stability not only appears to be a dimension of considerable emotional significance but also is evidently influenced by social and developmental forces.

The dimensions of clarity and consistency are equally neglected. There is every logical reason—and virtually no empirical evidence—to suggest that these dimensions (which no doubt overlap with the others) have important emotional consequences for the individual and will affect behavior. We have little doubt that if social influences bearing on these self-concept dimensions were sought, they would surely be found.

*Ego-Extensions.* At the very outset of his discussion of the "empirical me," James stressed the importance of those objects or entities physically external to the individual which nonetheless represented such important components of the self-concept. Some of these external entities—especially groups—have commanded considerable interest; others—especially significant others—have also been discussed at length; and still others, such as the products of our labor, or other physical objects, are often overlooked. Examples where external objects represent components of the self come readily to mind. A mother may take particular pride in her child, her husband in his shiny new automobile; a professional may take special pride in his academic degree, a businessman in his bank account or stock portfolio, a worker in his house or car. The failure of an owner's farm or small business may be experienced as a diminution or partial negation of the self. The significance of ego-extensions is also affected by age. External objects incorporated into the self of an adolescent are not those constituting a part of the child's self.

There is much to be learned in this area. Currently, little is known about the *degree* to which these various components are introjected; about the boundaries of the self, that is, the point beyond which the exterior component shifts from self to non-self; about the process by which these components are extruded or internalized; or about the interpersonal, contextual, and social structural factors that contribute to the introjection of these components. It seems apparent that the introjection of such external entities has important emotional consequences and is heavily influenced by contextual and social structural forces.

*Desired Selves.* One of the most exciting areas in need of additional research is the desired self. There is, to be sure, a very large body of research, extending back at least to the forties, on levels of aspiration and expectation—occupational, educational, economic—both for the self and for ego-extensions (e.g., Rosen, 1959). But the desired self includes much more.

Everyone has a vision, clear or vague, of the self—usually of a more beautiful, virtuous, admirable, and rewarding self. It may be a self that is located in a social setting; living in a large split-level; puttering in the garden; bearded, discussing Marx at 3 A.M. in Greenwich Village; well-dressed, dynamic, bright young executive moving swiftly ahead; sharing wine, cheese, and pipe smoke in discussions of metaphysics in a walnut-paneled book-lined study; and so on.

Although there is an interesting clinical literature, little systematic research is available. What is the content of these images? How clear or vague, well- or ill-defined, are they? To what extent are they pleasurable fantasies, to what extent serious commitments? How much do these imaginary or potential selves intrude on *current* behavior? In addition, the moral self—the self that one *should* be—is no less important, no more narrow in scope. Little is known about the broad range of shoulds that people impose on themselves—that they should know everything about wine, impress everyone favorably, dominate any situation, work 18 hours a day, always remain calm and unruffled, be an ideal father, and so on. These are demands we impose on ourselves and which assuredly affect our behavior and our mental health.

No one will argue that it is easy to study these regions of the self-concept systematically, but neither is it totally beyond our reach. Excellent work was done in this area over two decades ago by Havighurst and MacDonald (1955). They asked school children in New Zealand to write a brief essay on the type of person they would like to be like when they grew up. Interesting developmental trends appeared. "This study shows that the same developmental trend in the ideal self occurs in New Zealand children that has been observed in American children. This trend goes from an identification in early childhood with a parental figure to a stage in late adolescence which may be symbolized by an attractive, visible young adult or by an imaginary character who is a composite of desirable qualities. Intermediate in this trend is a stage of romanticism and glamor, when the ideal self is a glamorous, unreal character such as a movie star, a military figure, or a character in juvenile fiction who possesses superhuman qualities. Some children appear to omit this intermediate stage, while others pass through it quickly, and still others prolong it into late adolescence" (Havighurst and Macdonald, 1955: 273).

*Presenting Self.* The presenting self, of course, is more firmly rooted in specific situations than other self-concept regions, and should be studied accordingly (Goffman, 1956). Although methodological difficulties are to be expected, in principle there is no insuperable obstacle to the study of what type of person the individual seeks to appear as in social interaction and to what ends. A classification of the types of social situations that call forth different types of presenting selves would be an important sociological contribution, and an understanding of the social structural factors underlying chronic or cross-situational presenting selves would be equally so. We do not totally lack guidance in this area, having available such carefully controlled studies as those described by Gergen (1968) as well as the brilliantly innovative ideas on psychodrama and sociodrama introduced by Jacob Moreno over a generation ago. Though Moreno used the method largely for clinical and therapeutic purposes, its value for research has long been recognized.

WHAT CAN BE LEARNED

In sum, the burden of these observations is that there is a great deal more to the self-concept than self-esteem, and that most of it has been neglected in research. Today self-concept research seems to be spinning its wheels, unable to get untracked. All the important sociological variables that might be expected to account for it— age, race, social class, birth order, sex, ethnicity, religion—are found to explain remarkably little variance in it. The chief reason is that "it" —whether measured by the Piers-Harris Children's Self-Concept Scale, the Coopersmith Self-Esteem Inventory, the Fitts Tennessee Self-Concept Scale (Total Net Positive Score), the Butler-Haigh self-ideal self-discrepancy measure, the Rosenberg Self-Esteem scale, the Osgood semantic differential, the Gough and Heilbrun Adjective Check List—turns out to be *global* self-esteem. Global self-esteem is important—extremely important, perhaps the most important single item of information about the self-concept one would wish to obtain —but it is not all there is to the self-concept.

To offer some hint of the changes that might be effected by going beyond self-esteem, consider an example of considerable interest today, that of male and female self-concepts. Many people argue that society treats women as inferior and incompetent, and that women internalize these social definitions of their worth and tend to develop feelings of inferiority.

The empirical evidence in this regard is surprisingly inconsistent, with some studies (F. Rosenberg and Simmons, 1975; Bush, Simmons, Hutchinson, and Blyth, 1977–78) showing lower self-esteem among girls, and others (Rosenberg, 1965; Maccoby and Jacklin, 1974) showing little sex difference. But there are a number of other aspects of the self-concept which show definite sex differences. For example, F. Rosenberg and Simmons (1975) and Simmons and F. Rosenberg (1975), comparing boys' and girls' self-concepts by age, uncovered the following finding: that although the self-*esteem* of boys and girls differed only modestly, at adolescence girls showed considerably higher *instability* of self-concept—their ideas about themselves tended to change more quickly, to vary from day to day. Furthermore, girls showed strikingly *higher self-consciousness*, expressed in such reactions as feeling nervous about talking in front of others, feeling uneasy if someone watched them work, and thinking about other people's reactions to them at public gatherings.

Maccoby and Jacklin's (1974) careful and extensive summary of research on sex differences, discussed in chapter 1, also suggests that girls appear to have lower self-*confidence* than boys. Although this fact is sometimes taken as evidence of the damaged self-concepts of girls, it should be noted that it is not the girls' self-confidence that is unrealistically low but the boys' self-confidence that is unrealistically high. In other words, it is not so much that girls underestimate their abilities as that boys overestimate theirs. Who, then, has the damaged self-concept?

Nor are the self-*values* of boys and girls the same. In the New York State study, we found that, although boys and girls were both highly concerned with being well-liked by others, girls more consistently gave this characteristic top priority. They were more likely to stress values of interpersonal harmony and success (likeable; easy to get along with; friendly, sociable, and pleasant; well-liked by many different people); to stress kindly virtues (kindness and consideration, sympathy and understanding); and to value aesthetic appreciation (a refined person who shows good taste in things). Boys, on the other hand, were more likely to stress motoric values and physical courage, interpersonal dominance, freedom from naïveté, and versatility.

Although less information is available on adults, Mulford and

Salisbury (1964) have shown that there are differences in the salience of the social identity elements of men and women. In response to the "Who am I?" question, women with children were more likely spontaneously to define themselves as mothers and wives than men were to define themselves as fathers and husbands.

In sum, whatever the ultimate resolution of the question regarding male and female self-*esteem*, it is evident that male and female self-*concepts* differ in a number of ways.

The same point can be made with regard to the impact of other sociodemographic factors on the self-concept. Many sociological and psychological studies focus on the relationship of some denigrated social identity element—a low-status racial, religious, or ethnic group; a low social class; a social label or stigmatized status—to self-esteem. Generally speaking, these studies show surprisingly small relationships. But if the self-*esteem* of the privileged and disprivileged differs little, this does not mean that the self-*concepts* do not differ. The data in chapter 6 showed little difference in self-esteem between advantaged and disprivileged children, but this does not mean that minority group members may not have lesser self-*confidence* in mastering the problems of the world, given the more forbidding societal obstacles they face.

The final reason for taking account of these neglected aspects of the self-concept, we suggest, is that we will never understand self-esteem unless we go beyond self-esteem. The reason is plain. What does the fact that someone has a low opinion of his intelligence, neatness or tact tell us about his self-esteem? Very little, unless we know something of his self-values—how much he cares about his intelligence, neatness, or tact. What does the fact that someone has doubts about his ability to master certain tasks tell us about his self-esteem? Not much, unless we know the connection between his self-confidence and his self-esteem. What does the fact that someone is a homosexual, mental patient, delinquent or member of another stigmatized social status tell us about his self-esteem? Almost nothing, unless we know whether he has committed himself to that status or continues to struggle against it. What does the fact that certain minority group members absorb the general negative attitudes of the broader society toward their groups tell us about their self-esteem? Very little, unless we know how central or peripheral to the self this particular ego-extension is. What does the fact that someone wishes to be President, Babe Ruth or Albert Einstein tell us about self-esteem? Again, very little, unless we know whether these are playful fantasies or serious commitments. Or what can we tell about the self-esteem of people who act subordinate, inferior, or self-denigrating in relation to those who have the power to satisfy or frustrate their values? Very little, unless we know the extent to which the presenting self corresponds to the extant self. It thus seems apparent that we will never gain a true appreciation of self-esteem, let alone of the self-concept, until we go beyond—indeed, far beyond—self-esteem.

## NOTES

1. Some of the selectivity mechanisms involved in interpersonal interaction are discussed in Bachman and Secord (1962).

2. There are also other methods employed, such as attempting to impress others directly, to cultivate excellence, or to achieve success in various endeavors. Although the present discussion is confined to the selectivity mechanism, many other aspects of human behavior and human activity are directed toward the search for positive reflected appraisals in the interests of self-esteem protection and enhancement.

3. Congress, evidently more hard-working than the rest of us, some time ago gave up vacations (of about 50 days), but it allowed itself to go into "recess" (of about 50 days). Since recess was apparently too unproductive, it has given up this practice and now spends about 50 days in "district work periods."

4. James' (1890: 302) description of the initial stage is characteristically perceptive: "It is as if all that visited the mind had to stand an entrance-examination, and just show its face so as to be either approved or sent back."

# Appendix A

## Scales

### A–1 New York State Self-Esteem Scale (Rosenberg Self-Esteem)

The RSE is a 10-item Guttman scale with a Coefficient of Reproducibility of 92 percent and a Coefficient of Scalability of 72 percent. Respondents are asked to strongly agree, agree, disagree, or strongly disagree with the following items (asterisks represent low self-esteem responses):

| | | | | | |
|---|---|---|---|---|---|
| (1) | On the whole, I am satisfied with myself. | SA | A | D* | SD* |
| (2) | At times I think I am no good at all. | SA* | A* | D | SD |
| (3) | I feel that I have a number of good qualities. | SA | A | D* | SD* |
| (4) | I am able to do things as well as most other people. | SA | A | D* | SD* |
| (5) | I feel I do not have much to be proud of. | SA* | A* | D | SD |
| (6) | I certainly feel useless at times. | SA* | A* | D | SD |
| (7) | I feel that I'm a person of worth, at least on an equal plane with others. | SA | A | D* | SD* |
| (8) | I wish I could have more respect for myself. | SA* | A* | D | SD |
| (9) | All in all, I am inclined to feel that I am a failure. | SA* | A* | D | SD |
| (10) | I take a positive attitude toward myself. | SA | A | D* | SD* |

The scale is based on "contrived items" (Stouffer *et al.*, 1953), and yields a 7-point scale. Scale Item I is contrived from the combined responses to items 3, 7, and 9. If the respondent answers 2 out of 3 or 3 out of 3 positively, he receives a positive (that is, low self-esteem) score for Scale Item I.

Scale Item II is contrived from the combined responses to items 4 and 5. One out of 2 or 2 out of 2 positive responses are considered positive for Scale Item II.

Scale Items III, IV, and V are scored simply as positive or negative based on responses to items 1, 8, and 10.

Scale Item VI is contrived from the combined responses to items 2 and 6. One out of 2 or 2 out of 2 positive responses are considered positive.

The reproducibility and scalability coefficients suggest that the items have satisfactory internal reliability. Silber and Tippett (1965) showed a two-week test-retest reliability of $r = .85$, and Claire McCullough found a two-week test-retest reliability of $r = .88$; both used small college samples.

An examination of these items suggests that they have face validity. Although it is reasonable to question one or another item, it would appear that the items deal with a general favorable or unfavorable global self-attitude.

### CONSTRUCT VALIDITY

Construct validity has been examined in two ways. The first is the conformity of the measure with theoretical expectations (Cronbach and Meehl, 1955). There is theoretical reason to think that, if the scale measured self-esteem, it should be empirically related to (1) depressive affect (2) anxiety and (3) peer-group reputation.

*Depressive Affect.* The New York State study revealed a clear relationship between the self-esteem scale and a 6-item Guttman scale of depressive affect. Only 4 percent of those with the highest self-esteem scores compared with 80 percent of those with the lowest scores were rated as "highly depressed" ($r = .3008$). In addition, in a small sample (50) of "normal controls" who had volunteered for research purposes in a medical installation, those with low self-esteem scores were much more likely to be described by nurses as "often gloomy" and as "frequently disappointed."

*Anxiety.* The self-esteem scale bears a strong relationship to the psychophysiological indicators of anxiety which so effectively differentiated normal and neurotic soldiers in World War II (Stouffer *et al.*, 1950). Sixty-nine percent of those with the lowest self-esteem compared with only 19 percent of those with the highest reported a relatively large number of these psychophysiological symptoms of anxiety ($r = .4848$).

*Peer-Group Reputation.* In a special sample of 272 high school seniors in Maryland, those with high self-esteem were more likely to obtain high sociometric ratings from their peers, that is, to be selected as a leader by at least two of their classmates and to be judged by others as likely to be chosen as a leader (Rosenberg, 1965: p. 26). Among the above-mentioned normal controls, those with high self-esteem were more likely to be described by the nurses as "well thought of," "makes good impression," "often admired," and "respected by others." Finally, in the New York State study, those with high self-esteem were (1) more likely to participate in extracurricular activities (2) more likely to have been elected as an officer of a club or school organization, and (3) more likely to be a self-designated informal opinion leader (see chapter 5 for details on the relationship of self-esteem to leadership in the high school).

*Convergent and Discriminant Validity.* The convergent and discriminant validity of the RSE have been examined by Silber and Tippett (1965) and Tippett and Silber (1965) in accordance with the multitrait-multimethod framework of Campbell and Fiske (1959). This study of 44 college students measured two traits (global self-esteem and stability of self-concept) by means of four different methods: the RSE (Guttman scale), the Kelley Repertory Test (a self-ideal discrepancy test), the Heath self-image questionnaire (a sum of 20 items dealing with self- and social-ideal discrepancy), and a psychiatrist's rating.

One test of the adequacy of the RSE is whether it shows convergent validity with measures of the same concept based on different methods (monotrait-heteromethod correlations). The correlations of RSE to the self-ideal discrepancy score was $r = .67$; to the self-image questionnaire, $r = .83$; and to the psychiatrist's rating, $r = .56$. One criterion of discriminant validity is whether the monotrait-heteromethod correlations are higher than the heterotrait-monomethod correlations. Although the New York State RSE and stability of self-concept measures were both based on Guttman scales (heterotrait-monomethod), their correlation was .53, which was lower than the monotrait-heteromethod correlations. The other criterion of discriminant validity is whether the monotrait-heteromethod correlations exceed the heterotrait-heteromethod correlations. The correlations between RSE (a self-esteem measure) and measures of self-concept *stability*, assessed by the self-ideal measure, the self-image questionnaire, and the psychiatrist's rating were $r = .40$, $r = .34$, and $r = .21$, respectively—considerably lower than the correlations of self-esteem measured by different methods. In addition, the correlation of RSE to (1) stability of views of other people, and (2) stability of perceptual speed (heterotrait-heteromethod correlations) were close to zero (Tippett and Silber, 1965). Other evidence of convergent validity is Crandall's (1973) finding that the correlation of RSE (based on a 10-item score) and the Coopersmith Self-Esteem Inventory (Coopersmith, 1967) was .60. There is thus evidence of both convergent and discriminant validity for the RSE. The Silber and Tippett data suggest that methods factors may play some role, though probably not a major one.

*Factor analyses.* A number of studies (Kaplan and Pokorny, 1969; Kohn, 1969; Hensley and Roberts, 1976; Carmines and Zeller, 1974; Zeller and Carmines, 1976) have discovered two factors in these 10 items. It would lead us too far astray to discuss the different logical underpinnings of Guttman scales and factor analyses, but it is the case that items judged unidimensional by one method will not necessarily be so judged by another. There is, however, additional information of interest. Based on a sample of 340 high school students, Carmines and Zeller (1974) emerged with two factors— a "positive self-esteem" factor and a "negative self-esteem" factor. Subjects were assigned scores on each of these factors, and correlations were computed between each of these scores and 16 specific "criterion variables." These criterion variables covered three substantive areas: socioeconomic background (e.g., parent's education and occupation); psychological dispositions (e.g., locus of control, anomia, trust in people, intelligence); and social and political attitudes (e.g., political efficacy, political cynicism, understanding of democratic principles). It turns out that the relationship between positive self-esteem and these 16 criterion variables and between negative self-esteem and these criterion variables *are almost identical*. According to Carmines and Zeller (1974), the positive self-esteem and negative self-esteem factors "seem to tap the same rather than different dimensions, *for their correlations with the criterion variables are almost identical to one another in terms of direction, strength, and consistency*" (italics ours). The differences in the sizes of the correlations, using either score, were trivial.

Like Carmines and Zeller, Kohn (1969) and Kohn and Schooler (1969) also discovered two factors, called a "self-confidence" and "self-deprecation" factor. It is noteworthy, however, that their use of a Guttman scale based on the combined items consistently showed stronger correlations with theoretically related variables than either of the factors treated separately.

The RSE is frequently scored according to the Likert format and appears to yield results similar to those appearing when the Guttman procedure is used; separate validation, however, has not been conducted. Discussions of the strengths and weaknesses of the RSE appear in Rosenberg 1965: Chap. 2; Wylie, 1974: 180–189; Gordon, 1969; Crandall, 1973: 81–83, and Wells and Marwell, 1976.

## A–2 Baltimore Self-Esteem Scale (Rosenberg-Simmons Self-Esteem)

The RSSE is a 6-item Guttman scale with a Coefficient of Reproducibility of 90 percent, and a Coefficient of Scalability of 65 percent. The items are as follows (asterisks represent low self-esteem responses):

1. Everybody has some things about him which are good and some things about him which are bad. Are more of the things about you . . . Good; *Bad; or *Both about the same?
2. Another kid said, "I am no good." Do you ever feel like this? *Yes, No.
3. A kid told me: "There's a lot wrong with me." Do you ever feel like this? *Yes, No.
4. Another kid said: "I'm not much good at anything." Do you ever feel like this? *Yes, No.
5. Another kid said, "I think I am no good at all." Do you ever feel like this? *Yes, No.
6. How happy are you with the kind of person you are? Are you . . . Very happy with the kind of person you are; Pretty happy; *A little happy; or *Not at all happy.

This measure, administered to a sample of 798 black and white 6th and 7th grade children in 18 schools in a large Midwestern city

(Simmons, Brown, Bush, and Blyth, 1977) in the mid-1970s, showed a Coefficient of Reproducibility of 93 percent and a Coefficient of Scalability of 76 percent.

The fact that the items form a unidimensional scale for the sample as a whole, however, does not guarantee that it is unidimensional for varying subgroups in that sample. (An empirical investigation of this issue appears in Guterman, 1977.) In this regard, we have been particularly concerned with the variables of age and race. It is possible that, because of differences in cognitive development, responses of younger children might not reflect the same dimension as the responses of older children. Second, if it is true, as some writers claim, that there are subcultural differences between blacks and whites, then the danger would exist that the same responses might reflect different meanings for the two races. We have therefore examined the internal reliability of the RSSE for the age groups and races separately. The Coefficients of Reproducibility are as follows: among black children age 8–11, 90 percent; 12–14, 90 percent, and 15 or older, 91 percent; among white children, the corresponding figures are 91 percent, 90 percent, and 90 percent. The internal reliability of the scale (based on Guttman criteria) is thus virtually identical for each age group and for both races.

## A–3   Stability of Self Scale (New York State)

The Stability of Self Scale (New York State) is a 5-item Guttman scale with a Coefficient of Reproducibility of 94 percent and a Coefficient of Scalability of 77 percent. The items are as follows (asterisks represent high stability responses):

1. Does your opinion of yourself tend to change a good deal or does it always continue to remain the same? Changes a good deal; *Changes somewhat; *Changes very little; *Does not change at all.

2. Do you ever find that on one day you have one opinion of yourself and on another day you have a different opinion? Yes, this happens often; *Yes, this happens sometimes; *Yes, this rarely happens; *No, this never happens.

3. I have noticed that my ideas about myself seem to change very quickly. Agree, *Disagree.

4. Some days I have a very good opinion of myself; other days I have a very poor opinion of myself. Agree, *Disagree.

5. I feel that nothing, or almost nothing, can change the opinion I currently hold of myself. *Agree, Disagree.

## A–4  Stability of Self Scale (Baltimore)

The Stability of Self Scale (Baltimore) is a 7-item Guttman scale with a Coefficient of Reproducibility of 89 percent and a Coefficient of Scalability of 65 percent. The items are as follows (asterisks represent high stability responses):

1. How sure are you that you know what kind of person you really are? Are you . . . *Very sure; *Pretty sure; Not very sure; or Not at all sure.

2. How often do you feel mixed up about yourself, about what you are really like? Often; Sometimes; or *Never.

3. Do you feel like this: "I know just what I'm like. I'm really sure about it." *Yes, No.

4. A kid told me: "Some days I like the way I am. Some days I do not like the way I am." Do your feelings *change* like this? Yes, *No.

5. A kid told me: "Some days I am happy with the kind of person I am, other days I am not happy with the kind of person I am." Do your feelings change like this? Yes, *No.

6. Do you . . . *Know for sure how nice a person you are; or, Do your ideas about how nice you are change a lot?

7. A kid told me: "Some days I think I am one kind of person, other days a different kind of person." Do your feelings *change* like this? Yes, *No.

## A–5  Depressive Affect Scale (Baltimore)

The Depressive Affect Scale (Baltimore) is a 6-item Guttman scale with a Coefficient of Reproducibility of 92 percent and a Coefficient of Scalability of 70 percent. The items are as follows (asterisks represent high depression):

1. How happy would you say you are most of the time? Would you say you are . . . Very happy; Pretty happy; Not very happy; or *Not at all happy.
2. Would you say this: "I get a lot of fun out of life." Yes, *No.
3. Would you say this: "Mostly, I think I am quite a happy person." Yes, *No.
4. How happy are you today: Are you . . . Very happy; Pretty happy; Not very happy; or *Not at all happy.
5. A kid told me: "Other kids seem happier than I." Is this . . . *True for you; or, Not true for you.
6. Would you say that most of the time you are . . . Very cheerful; Pretty cheerful; Not very cheerful; or *Not cheerful at all.

## A–6  Self-Consciousness Scale

The Self-Consciousness Scale is a 7-item Guttman scale with a Coefficient of Reproducibility of 89 percent and a 63 percent Coefficient of Scalability. The items are as follows (asterisks represent high self-consciousness responses):

1. Let's say some grownup or adult visitor came into class and the teacher wanted them to know who you were, so she asked you to stand up and tell him a little about yourself. Would you like that; *Would you not like it; or Wouldn't you care?
2. If the teacher asked you to get up in front of the class and talk a little bit about your summer, would you be . . . *Very nervous; a Little nervous; or Not at all nervous?
3. If you did get up in front of the class and tell them about your summer, . . . *Would you think a lot about how all the kids were looking at you; Would you think a little bit about how all the kids were looking at you; or Wouldn't you think at all about the kids looking at you?
4. If you were to wear the wrong kind of clothes to a party, would that bother you . . . *A lot; A little; or Not at all?
5. If you went to a party where you did not know most of the kids, would you wonder what they were thinking about you? *Yes, No.
6. Do you get nervous when someone watches you work? *Yes, No.
7. A young person told me: "When I'm with people I get nervous because I worry how much they like me." Do you feel like this . . . *Often; Sometimes; or Never.

# Appendix B

## Indicators

### B–1 Perceived Self-Concept Indicators

The following items served as indicators of the perceived self:

1. Would you say your mother thinks you are: A wonderful person; A pretty nice person; A little bit of a nice person; or Not such a nice person. (Same question asked about "your father," "your teachers," and "kids in your class.")

2. How much do boys like you? Very much; Pretty much; Not very much; Not at all.

3. How much do girls like you? Very much; Pretty much; Not very much; Not at all.

4. Let's pretend your parents wanted to tell someone all about you. What type of person would they say you are? (The description was coded into the following three categories: only favorable and neutral remarks; remarks either all neutral or both favorable and unfavorable; only unfavorable and neutral remarks.) The same question was asked regarding "your teachers" and "your best friend," and the same coding categories applied.

## B–2   Racial Group Disidentification

1. If someone said something bad about the Negro or colored race, would you feel almost as if they had said something bad about you? Yes, No.
2. Is being Negro or colored very important to you, pretty important, or not important to you?
3. Do you feel very proud of being Negro, pretty proud, not very proud, or not at all proud?
4. If you could be anything in the world when you grow up, would you want to be Negro, white, or something else?
5. If you could be born again, would you like to be born of a different race, not Negro or colored? Yes, Maybe, No.
6. Do you think you would be happier if you were not Negro? Yes, Maybe, No.

These questions were asked only of black children. The term "Negro" was used because the interviews were conducted in 1968. Today, of course, the term "black" would be used.

## B–3   Religious Group Disidentification

1. If someone said something bad about your religion, would you almost feel if they had said something bad about you? Yes, No.
2. Is your religious group very important to you, pretty important, or not very important?
3. Do you feel very proud of your religious group, pretty proud, not very proud, or not at all proud?

# REFERENCES

The references below do not include sources cited only within quotations from other sources.

Allport, G. W. 1943. The ego in contemporary psychology. *Psychological Review* 50: 451–478.
———. 1954. The historical background of modern social psychology. In G. Lindzey, ed. *Handbook of Social Psychology*. Cambridge, Massachusetts: Addison-Wesley. Vol. I.
———. 1955. *Becoming*. New Haven: Yale University Press.
———. 1961. *Pattern and Growth in Personality*. New York: Holt, Rinehart and Winston.
Allport, G. W., and Odbert, H. S. 1936. Trait-names: a psycho-lexical study. *Psychological Monographs*, No. 211.
Angyal, A. 1941. *Foundations for a Science of Personality*. New York: Commonwealth Fund.
Anisfeld, M., Bogo, N., and Lambert, W. E. 1962. Evaluational reactions to accented English speech. *Journal of Abnormal and Social Psychology* 65: 223–231.
Ausubel, D. P. 1958. Ego development among segregated Negro children. *Mental Hygiene* 42: 362–369.
Bachman, J. G. 1970. *Youth in Transition, Volume II: The Impact of Family Background and Intelligence on Tenth-Grade Boys*. Ann Arbor, Michigan: Survey Research Center, Institute for Social Research.
Backman, C. W., and Secord, P. F. 1959. The effect of perceived liking on interpersonal attraction. *Human Relations* 12: 379–384.
———. 1962. Liking, selective interaction, and misperception in congruent interpersonal relations. *Sociometry* 25: 321–335.
Backman, C. W., Secord, P. F., and Pierce, J. R. 1963. Resistance to change in the self-concept as a function of consensus among significant others. *Sociometry* 26: 102–111.
Bakan, D. 1971. Adolescence in America: from idea to social fact. *Daedalus* 100: 979–995.
Baldwin, J. M. 1897. *Social and Ethical Interpretations in Mental Development*. New York: Macmillan.
Barton, A. H. 1970. Comment on Hauser's "Context and consex." *American Journal of Sociology* 76: 514–517.
Baughman, E. 1971. *Black Americans: A Psychological Analysis*. New York and London: Academic Press.
Beck, A. T. 1967. *Depression: Clinical, Experimental and Theoretical Aspects*. New York: Hoeber.
Beers, J. S. 1973. Self-esteem of black and white fifth-grade pupils as a function of demographic categorization. Paper presented at the annual meeting of the American Educational Research Association, New Orleans, February.
Bem, D. J. 1965. An experimental analysis of self-persuasion. *Journal of Experimental Social Psychology* 1: 199–218.
———. 1967. Self-perception: an alternative interpretation of cognitive dissonance phenomena. *Psychological Review* 74: 183–200.

# References

———. 1972. *Beliefs, Attitudes and Human Affairs*. Belmont, California: Brooks Cole.

Berger, P. L., and Luckmann, T. 1966. *The Social Construction of Reality*. New York: Doubleday and Company.

Blalock, H. M. 1961. Evaluating the relative importance of variables. *American Sociological Review* 26: 866–874.

Blau, P. M. 1960. Structural effects. *American Sociological Review* 25: 178–193.

Blos, P. 1962. *On Adolescence: A Psychoanalytic Interpretation*. New York: The Free Press.

Bogardus, E. S. 1928. *Immigration and Race Attitudes*. Boston: Heath.

———. 1959. Race reactions by sexes. *Sociology and Social Research* 43: 439–441.

Brewer, M. B., and Campbell, D. T. 1976. *Ethnocentrism and Intergroup Attitudes: East African Evidence*. New York: Halsted.

Brigham, J. C. 1974. Views of black and white children concerning the distribution of personality characteristics. *Journal of Personality* 42: 144–158.

Brim, O. G. 1974. The sense of personal control over one's life. Invited address to Divisions 7 and 8 of the American Psychological Association, New Orleans, September.

Brissett, D. 1972. Toward a clarification of self-esteem. *Psychiatry* 35: 255–262.

Brody, E. B. 1964. Color and identity conflict in young boys. *Archives of General Psychiatry* 10: 354–360.

Brookover, W. B., Thomas, S., and Paterson, A. 1964. Self-concept of ability and school achievement. *Sociology of Education* 37: 271–278.

Bush, D., Simmons, R., Hutchinson, B., and Blyth, D. 1977–78. Adolescent perception of sex roles in 1968 and 1975. *Public Opinion Quarterly* 41: 459–474.

Butler, J. M., and Haigh, G. V. 1954. Changes in the relation between self-concepts and ideal concepts consequent upon client-centered counseling. In C. R. Rogers and R. F. Dymond, eds. *Psychotherapy and Personality Change*. Chicago: University of Chicago Press. Pp. 55–75.

Campbell, D. T., and Fiske, D. W. 1959. Convergent and discriminant validation by the multitrait-multimethod matrix. *Psychological Bulletin* 56: 81–105.

Campbell, E. Q., and Alexander, C. N. 1965. Structural effects and interpersonal relationships. *American Journal of Sociology* 71: 284–289.

Campbell, J. D. 1967. Studies in attitude formation: the development of health orientations. In C. W. Sherif and M. Sherif, eds. *Attitude, Ego-Involvement and Change*. New York: Wiley. Pp. 7–25.

———. 1975. Illness is a point of view: the development of children's concepts of illness. *Child Development* 46: 92–100.

Carmines, E. G., and Zeller, R. A. 1974. On establishing the empirical dimensionality of theoretical terms: an analytical example. *Political Methodology* 1: 75–96.

Carter, T. P. 1968. Negative self-concepts of Mexican-American students. *School and Society* 96: 217–219.

Cartwright, D. 1950. Emotional dimensions of group life. In M. L. Reymert, ed. *International Symposium on Feelings and Emotions*. New York: McGraw-Hill. Pp. 439–447.

Chein, I. 1944. The awareness of self and the structure of the ego. *Psychological Review* 51: 304–314.

Christmas, J. 1973. Self-concept and attitudes. In R. M. Dreger and K. S. Miller, eds. *Comparative Studies of Blacks and Whites in the United States*. New York: Seminar Press. Pp. 249–272.

Clark, K. B. 1965. *Dark Ghetto: Dilemmas of Social Power*. New York: Harper and Row.

Clark, K. B., and Clark, M. P. 1947. Racial identification and preference in Negro children. In T. M. Newcomb and E. L. Hartley, eds. *Readings in Social Psychology*. New York: Holt. Pp. 169–178.

Clark, T. P. 1970. *Mexican-Americans in School: A History of Educational Neglect*. New York: College Entrance Examination Board.

Coleman, J. 1961. *The Adolescent Society*. New York: The Free Press.

Coleman, J. S., Campbell, E. Q., Hobson, C. J., McPartland, J., Mood, A. M., Weinfeld, F. D., and York, R. L. 1966. *Equality of Educational Opportunity*. Office of

Education, U.S. Department of Health, Education and Welfare. Washington, D.C.: U.S. Government Printing Office.

Combs, A. W., and Snygg, D. 1959. *Individual Behavior*. Revised Edition. New York: Harper.

Cooley, C. H. 1912. *Human Nature and the Social Order*. New York: Scribners.

Cooper, J. G. 1971. Perception of self and others as a function of ethnic group membership. *Eric*: ED 057 965.

Coopersmith, S. 1967. *The Antecedents of Self-Esteem*. San Francisco: Freeman.

Crain, R. L., and Weissman, C. S. 1972. *Discrimination, Personality and Achievement: A Survey of Northern Blacks*. New York: Seminar Press.

Crandall, R. 1973. The measurement of self-esteem and related constructs. In J. P. Robinson and P. R. Shaver, eds. *Measures of Social Psychological Attitudes: Revised Edition*. Ann Arbor, Michigan: Institute for Social Research. Pp. 45–168.

Cronbach, L. J., and Meehl, P. E. 1955. Construct validity in psychological tests. *Psychological Bulletin* 52: 177–193.

Crowne, D. P., and Marlowe, D. 1964. *The Approval Motive*. New York: Wiley.

Davis, A., Gardner, B. B., and Gardner, M. 1944. *Deep South: A Social Anthropological Study of Caste and Class*. Chicago: University of Chicago Press.

Davis, J. A. 1966. The campus as a frog pond. *American Journal of Sociology* 72: 17–31.

———. 1971. *Elementary Survey Analysis*. Englewood Cliffs, New Jersey: Prentice-Hall.

Davis, K. 1944. Adolescence and the social structure. *Annals of the American Academy of Political and Social Science* 236: 8–16.

Denzin, N. K. 1966. The significant others of a college population. *Sociological Quarterly* 7: 298–310.

Deutsch, M., and Solomon, L. 1959. Reactions to evaluations by others as influenced by self-evaluations. *Sociometry* 22: 93–112.

Dittes, J. E. 1959. Attractiveness of group as a function of self-esteem and acceptance by group. *Journal of Abnormal and Social Psychology* 59: 77–82.

Douvan, E., and Adelson, J. 1966. *The Adolescent Experience*. New York: Wiley.

Durkheim, E. 1947. *The Division of Labor in Society*. Glencoe, Illinois: The Free Press.

———. 1951. *Suicide*. Glencoe, Illinois: The Free Press.

Eagly, A. H. 1967. Involvement as a determinant of response to favorable and unfavorable information. *Journal of Personality and Social Psychology* Monograph 7, no. 3, pt. 2.

Edelson, M., and Jones, A. E. 1954. Operational explorations of the conceptual self system and the interaction between frames of reference. *Genetic Psychology Monographs* 50: 43–139.

Engel, M. 1959. The stability of the self-concept in adolescence. *Journal of Abnormal and Social Psychology* 58: 211–215.

Epps, E. C. 1969. Correlates of academic achievement among Northern and Southern urban Negro students. *Journal of Social Issues* 25: 55–70.

Epstein, S. 1973. The self-concept revisited: or a theory of a theory. *American Psychologist* 28: 404–416.

Erdelyi, M. H. 1974. A new look at the new look: perceptual defense and vigilance. *Psychological Review* 81: 1–25.

Eriksen, B. A., and Eriksen, C. W. 1972. *Perception and personality*. Morristown, New Jersey: General Learning Press.

Erikson, E. H. 1956. The problem of ego-identity. *Journal of the American Psychoanalytic Association* 4: 56–121.

———. 1959. Identity and the life cycle: selected papers. *Psychological Issues* 1: 1–171.

———. 1966. The concept of identity in race relations: notes and queries. *Daedalus* 95: 145–171.

Farley, R., and Hermalin, A. I. 1971. Family stability: a comparison of trends between blacks and whites. *American Sociological Review* 36: 1–17.

Festinger, L. 1954. A theory of social comparison processes. *Human Relations* 7: 117–140.

Fisher, S., and Cleveland, S. E. 1958. *Body Image and Personality*. Princeton: Van Nostrand.

# References

Fitch, G. 1970. Effects of self-esteem, perceived performance, and choice on causal attributions. *Journal of Personality and Social Psychology* 16: 311–315.

Flapan, D. 1968. *Children's Understanding of Social Interaction*. New York: Teachers College.

Flavell, J. 1974. The development of inferences about others. In T. Mischel, ed. *Understanding Other Persons*. Oxford: Blackwell.

Franks, D., and Marolla, J. 1976. Efficacious action and social approval as interacting dimensions of self-esteem: formulation through construct validation. *Sociometry* 39: 324–341.

French, J. R. P. 1968. The conceptualization of mental health in terms of self-identity theory. In S. B. Sells, ed. *The Definition and Measurement of Mental Health*. Washington, D.C.: Department of Health, Education, and Welfare.

Freud, A. 1946. *The Ego and the Mechanisms of Defense*. New York: International Universities Press.

———. 1958. Adolescence. *Psychoanalytic Study of the Child* 13: 255–278.

Freud, S. 1933. *A New Series of Introductory Lectures on Psycho-Analysis* (Trans. W. J. H. Sprott). New York: Norton.

Fromm, E. 1941. *Escape from Freedom*. New York: Rinehart.

———. 1947. *Man for Himself*. New York: Rinehart.

Gergen, K. 1968. Personal consistency and the presentation of self. In C. Gordon and K. Gergen, eds. *The Self in Social Interaction*. New York: Wiley. Pp. 299–308.

———. 1971. *The Concept of Self*. New York: Holt, Rinehart and Winston.

Gerth, H. H., and Mills, C. W. 1946. *From Max Weber: Essays in Sociology*. New York: Oxford.

Gill, B. 1975. *Here at the New Yorker*. New York: Random House.

Gillman, G. B. 1969. The relationship between self-concept, intellectual ability, achievement and manifest anxiety among select groups of Spanish surname migrant students in New Mexico. *Eric*: Ed 029 723.

Goffman, E. 1955. On face-work: an analysis of ritual elements in social interaction. *Psychiatry: Journal for the Study of Interpersonal Processes* 18: 213–231.

———. 1956. *The Presentation of Self in Everyday Life*. Edinburgh: University of Edinburgh.

———. 1963. *Stigma: Notes on the Management of Spoiled Identity*. New York: Prentice-Hall.

Goodman, M. E. 1952. *Race Awareness in Young Children*. Cambridge, Mass.: Addison-Wesley.

Gordon, C. 1963. *Self-Conception and Social Achievement*. Unpublished Ph.D. dissertation, University of California at Los Angeles. Ann Arbor: University Microfilms.

———. 1968. Self-conceptions: configurations of content. In C. Gordon and K. J. Gergen, eds. *The Self in Social Interaction*. New York: Wiley. Pp. 115–136.

———. 1971. Social characteristics of early adolescence. *Daedalus* 100: 931–960.

———. 1974. Persons-conceptions analytic dictionary for the General Inquirer System of computer-aided content analysis. Unpublished paper.

———. 1976. Development of evaluated role identities. *Annual Review of Sociology* 2: 405–433.

Gough, H. G., and Heilbrun, A. B. 1965. *The Adjective Check List Manual*. Palo Alto, California: Consulting Psychologists Press.

Gray, J. S., and Thompson, A. H. 1953. The ethnic prejudices of white and Negro college students. *Journal of Abnormal and Social Psychology* 48: 311–313.

Grier, W. H., and Cobbs, P. M. 1968. *Black Rage*. New York: Basic Books.

Grinker, R. R., Sr., Grinker, R. R., Jr., and Timberlake, J. 1962. A study of "mentally healthy" young males (homoclites). *American Medical Association Archives of General Psychiatry* 6: 405–453.

Guterman, S. S. 1972. *Black Psyche: The Modal Personality Patterns of Black Americans*. Berkeley, California: Glendessary Press.

———. 1977. Are personality measures equally valid within different social classes? A case study using locus of control. Paper presented at the 72nd Annual Meeting of the American Sociological Association, Chicago, Ill.: September.

Guttentag, M. 1970. Group cohesiveness, ethnic organization, and poverty. *Journal of Social Issues* 26: 105–132.

Guzman, L. P. 1976. *Puerto Rican Self-Esteem and the Importance of Social Context.* Unpublished Ph.D. dissertation, State University of New York at Buffalo.

Hall, C. S., and Lindzey, G. 1957. *Theories of Personality.* New York: Wiley.

Hall, G. S. 1904. *Adolescence.* New York: Appleton.

Harding, J., Proshansky, H., Kutner, B., and Chein, I. 1969. Prejudice and ethnic relations. In G. Lindzey and E. Aronson, eds. *Handbook of Social Psychology,* Vol. V. Reading, Massachusetts: Addison-Wesley. Pp. 1–76.

Harris, Louis and Associates. 1975. *The Myth and Reality of Aging in America.* Washington, D.C.: The National Council on the Aging.

Hauser R. M. 1970. Context and consex: a cautionary tale. *American Journal of Sociology* 75: 645–664.

Havighurst, R., and MacDonald, D. V. 1955. Development of the ideal self in New Zealand and American children. *Journal of Educational Research* 49: 263–276.

Healey, G. W., and DeBlassie, R. R. 1974. A comparison of Negro, Anglo and Spanish American adolescents' self-concept. *Adolescence* 9: 15–24.

Heider, F. 1958. *The Psychology of Interpersonal Relations.* New York: Wiley.

Heiss, J., and Owens, S. 1972. Self-evaluations of blacks and whites. *American Journal of Sociology* 78: 360–370.

Hensley, W. E., and Roberts, M. K. 1976. Dimensions of Rosenberg's self-esteem scale. *Psychological Reports* 38: 583–584.

Herman, S. N. 1970a. *Israelis and Jews: The Continuity of an Identity.* New York: Random House.

———. 1970b. *American Students in Israel.* Ithaca, New York: Cornell University Press.

Hessler, R. M., New, P. K., Kubish, P., Ellison, D. L., and Taylor, F. H. 1971. Demographic context, social interaction, and perceived health status: Excedrin headache #1. *Journal of Health and Social Behavior* 12: 191–199.

Hilgard E. R. 1949. Human motives and the concept of the self. *American Psychologist* 4: 374–382.

Hishiki, P. C. 1969. Self-concept of sixth grade girls of Mexican descent. *California Journal of Educational Research* 20: 56–62.

Hobbes, T. 1887. *Leviathan.* London: George Routledge and Sons.

———. 1949. On the natural condition of mankind as concerning their felicity and misery. In L. Wilson and W. L. Kolb, eds. *Sociological Analysis.* New York: Harcourt Brace (Originally published in *Leviathan,* Part I, Chapter XIII, 1651). Pp. 160–163.

Hodge, R. W., Siegel, P. M., and Rossi, P. H. 1964. Occupational prestige in the U.S., 1925–1963. *American Journal of Sociology* 70: 286–302.

Hollingshead, A. B., and Redlich, F. C. 1958. *Social Class and Mental Illness.* New York: Wiley.

Homans, G. C. 1958. Social behavior as exchange. *American Journal of Sociology* 63: 597–606.

Horney, K. 1945. *Our Inner Conflicts.* New York: Norton.

———. 1950. *Neurosis and Human Growth.* New York: Norton.

Horowitz, E. L. 1944. "Race" attitudes. In O. Klineberg, ed. *Characteristics of the American Negro.* New York: Harper. Pp. 139–247.

Hughes, E. C. 1962. What other? In A. M. Rose, ed. *Human Behavior and Social Processes.* Boston: Houghton Mifflin. Pp. 119–127.

Hunt, D. E., and Hardt, R. H. 1969. The effects of Upward Bound Programs on the attitudes, motivation and academic achievements of Negro students. *Journal of Social Issues* 25: 122–124.

Hunt, L. L., and Hunt, J. G. 1975. A religious factor in secular achievement among blacks: the case of Catholicism. *Social Forces* 53: 595–606.

Jahoda, M. 1958. *Current Concepts of Positive Mental Health.* New York: Basic Books.

Jahoda, M., Lazarsfeld, P. F., and Zeisel, H. 1971. *Marienthal: The Sociography of an Unemployed Community.* Chicago: Aldine.

James, W. 1890. *The Principles of Psychology.* Reprint. New York: Dover, 1950.

# References

Jensen, G. F. 1972. Delinquency and adolescent self-conceptions: a study of the personal relevance of infraction. *Social Problems* 20: 84–103.

Johnson, T. J., Feigenbaum, R., and Weiby, M. 1964. Some determinants and consequences of the teacher's perception of causation. *Journal of Educational Psychology* 55: 237–246.

Jones, E. E., Gergen, K. J., and Davis, K. E. 1962. Some determinants of reactions to being approved or disapproved as a person. *Psychological Monographs* 72: 1–17.

Jones, E. E., and Davis, K. E. 1965. From acts to dispositions. In L. Berkowitz, ed. *Advances in Experimental Social Psychology*. Vol. 2. New York: Academic Press. Pp. 219–266.

Jones, S. C. 1973. Self and interpersonal evaluations: esteem theories versus consistency theories. *Psychological Bulletin* 79: 185–199.

Kaplan, H. B. 1971. Social class and self-derogation: a conditional relationship. *Sociometry* 34: 41–65.

———. 1975. *Self-Attitudes and Deviant Behavior*. Pacific Palisades, California: Goodyear Publishing.

Kaplan, H. B., and Pokorny, A. D. 1969. Self-derogation and psycho-social adjustment. *Journal of Nervous and Mental Disease* 149: 421–434.

Kardiner, A., and Ovesey, L. 1951. *The Mark of Oppression*. New York: Norton.

Karweit, N. 1977. Issues in the Measurement of Contextual Effects: Homogeneity of Associations and Multiple Reference Populations. Center for Social Organization of Schools, Johns Hopkins, Report No. 224.

Katz, D., and Braly, K. W. 1935. Racial prejudice and racial stereotypes. *Journal of Abnormal and Social Psychology* 30: 175–193.

Kelley, H. H. 1952. Two functions of reference groups. In G. E. Swanson, T. M. Newcomb, and E. L. Hartley, eds. *Readings in Social Psychology*. Revised Edition. New York: Holt. Pp. 410–414.

———. 1967. Attribution theory in social psychology. In D. Levine, ed. *Nebraska Symposium on Motivation*. Omaha: University of Nebraska Press. Pp. 192–238.

Kinch, J. W. 1963. A formalized theory of the self-concept. *American Journal of Sociology* 68: 481–486.

———. 1967. Experiments on factors related to self-concept change. In J. G. Manis and B. N. Meltzer, eds. *Symbolic Interaction*. Boston: Allyn and Bacon. Pp. 262–268.

Kohlberg, L., and Gilligan, C. 1971. The adolescent as a philosopher: the discovery of the self in a postconventional world. *Daedalus* 100: 1051–1086.

Kohn, M. L. 1969. *Class and Conformity: A Study in Values*. Homewood, Illinois: Dorsey Press.

Kohn, M. L., and Schooler, C. 1969. Class, occupation, and orientation. *American Sociological Review* 34: 659–678.

Komarovsky, M. 1946. Cultural contradictions and sex roles. *American Journal of Sociology* 52: 184–189.

Krech, D., and Crutchfield, R. S. 1948. *Theory and Problems of Social Psychology*. New York: McGraw-Hill.

Kuhn, M. H. 1960. Self-attitudes by age, sex and professional training. *Sociological Quarterly* 9: 39–55.

———. 1964. The reference group reconsidered. *Sociological Quarterly* 5: 5–24.

Kuhn, M. H., and McPartland, T. 1954. An empirical investigation of self-attitudes. *American Sociological Review* 19: 68–76.

Kvaraceus, W. C. et al. 1965. *Negro Self-Concept: Implications for School and Citizenship*. New York: McGraw-Hill.

Laurence, J. E. 1970. White socialization: black reality. *Psychiatry* 33: 174–194.

Lazarsfeld, P. F., and Thielens, W. 1958. The social context of apprehension. *The Academic Mind*. New York: The Free Press. Pp. 237–265.

Lecky, P. 1945. *Self-Consistency: A Theory of Personality*. New York: Island Press.

Lewin, K. 1948. *Resolving Social Conflicts*. New York: Harper.

———. 1951. *Field Theory in Social Science*. New York: Harper.

Lively, W. J., and Bromley, D. B. 1973. *Person Perception in Childhood and Adolescence*. London: Wiley.

Luck, P. W., and Heiss, J. 1972. Social determinants of self-esteem in adult males. *Sociology and Social Research* 57: 69–84.

Lynd, H. M. 1958. *On Shame and the Search for Identity*. New York: Harcourt Brace.

Maccoby, E. E., and Jacklin, C. N. 1974. *The Psychology of Sex Differences*. Stanford, California: Stanford University Press.

Maehr, M. L., Mensing, J., and Nafzger, S. 1962. Concept of self and reactions of others. *Sociometry* 25: 353–357.

Manis, M. 1955. Social interaction and the self-concept. *Journal of Abnormal and Social Psychology* 51: 362–370.

Maslow, A. H. 1954. *Motivation and Personality*. New York: Harper.

———. 1956. Personality problems and personality growth. In C. E. Moustakas, ed. *The Self: Explorations in Personal Growth*. New York: Harper.

McCall, G. J., and Simmons, J. L. 1966. *Identities and Interactions*. New York: The Free Press.

McCarthy, J. D., and Yancey, W. L. 1971. Uncle Tom and Mr. Charlie: Metaphysical pathos in the study of racism and personal disorganization. *American Journal of Sociology* 76: 648–672.

McDaniel, E. L. 1967. Relationship between self-concept and specific variables in a low income culturally different population. Final report on Head Start evaluation and research. *Eric:* ED 019 124.

McDill, E. L., Meyers, E. D., Jr., and Rigsby, L. C. 1966. *Sources of Educational Climate in High School*. Department of Social Relations, Johns Hopkins University.

McDonald, R. L., and Gynther, M. D. 1965. Relationship of self and ideal-self descriptions with sex, race and class of southern adolescents. *Journal of Personality and Social Psychology* 1: 85–88.

McDougall, W. 1932. *The Energies of Men*. London: Methuen.

McGuire, W. J. et al. 1978. Salience of ethnicity in the spontaneous self-concept as a function of one's ethnic distinctiveness in the social environment. *Journal of Personality and Social Psychology* 36: 511–520.

Mead, G. H. 1934. *Mind, Self and Society*. Chicago: University of Chicago Press.

Meltzer, H. 1939a. Group differences in nationality and race preferences of children. *Sociometry* 2: 86–105.

———. 1939b. Nationality preferences and stereotypes of colored children. *Journal of Genetic Psychology* 54: 403–424.

Merton, R. K. 1957. *Social Theory and Social Structure*. Revised edition. New York: The Free Press.

Merton, R. K., and Kitt, A. S. 1950. Contributions to the theory of reference group behavior. In R. K. Merton and P. F. Lazarsfeld, eds. *Continuities in Social Research: Studies in the Scope and Method of "The American Soldier."* Glencoe, Illinois: The Free Press. Pp. 40–105.

Meyer, J. W. 1970. High school effects on college intentions. *American Journal of Sociology* 76: 59–70.

Middleton, R. 1972. Self-esteem and psychological impairment among American black, American white, and African men: preliminary report. Mimeographed.

———. 1977. Personal Communication. April.

Miller, D. T., and Ross, M. 1975. Self-serving biases in the attribution of causality: fact or fiction? *Psychological Bulletin* 82: 213–225.

Milner, E. 1953. Some hypotheses concerning the influence of segregation on Negro personality development. *Psychiatry* 16: 291–297.

Miyamoto, S. F., and Dornbusch, S. 1956. A test of the symbolic interactionist hypothesis of self-conception. *American Journal of Sociology* 61: 399–403.

Mordock, J. B. 1975. *The Other Children: An Introduction to Exceptionality*. New York: Harper and Row.

Moustakas, C. E., ed. 1956. *The Self: Explorations in Personal Growth*. New York: Harper.

Mulford, H. A., and Salisbury, W. W. 1964. Self-conceptions in a general population. *Sociological Quarterly* 5: 35–46.

Murphy, G. 1947. *Personality*. New York: Harper.

# References

Najmi, M. A. K. 1962. Comparison of Greeley's Spanish-American and Anglo-white elementary school children's responses to instruments designed to measure self-concept and some related variables. Unpublished doctoral dissertation, Colorado State University. Microfilms, Ann Arbor.

Nelson, J. I. 1972. High school context and college plans: the impact of social structure on aspirations. *American Sociological Review* 37: 143–148.

Newcomb, T. 1950. *Social Psychology.* New York: Dryden.

North, C. C., and Hatt, P. K. 1953. Jobs and occupations: a popular evaluation. In R. Bendix and S. M. Lipset, eds. *Class, Status and Power.* Glencoe, Illinois: The Free Press. Pp. 411–426.

Offer, D. 1969. *The Psychological World of the Teen-Ager.* New York: Basic Books.

Osgood, C. E., Suci, G. J., and Tannenbaum, P. H. 1957. *The Measurement of Meaning.* Urbana, Illinois: University of Illinois Press.

Park, R. E. 1928. Human migration and the marginal man. *American Journal of Sociology* 33: 881–893.

Peterson, B., and Ramirez, M., III. 1971. Real-ideal self disparity in Negro and Mexican-American children. *Psychology* 8: 22–28.

Pettigrew, T. F. 1964. *A Profile of the Negro American.* Princeton, New Jersey: Van Nostrand.

———. 1967. Social evaluation theory: convergences and applications. In D. Levine, ed. *Nebraska Symposium on Motivation, 1967.* Lincoln, Nebraska: University of Nebraska Press. Pp. 241–311.

Piaget, J. 1928. *Judgment and Reasoning in the Child.* London: Routledge and Kegan Paul.

———. 1929. *The Child's Conception of the World.* London: Routledge and Kegan Paul.

———. 1932. *The Language and Thought of the Child.* London: Routledge and Kegan Paul.

———. 1948. *The Moral Judgment of the Child.* Glencoe, Illinois: The Free Press.

Piers, E. V. 1969. *Manual for the Piers-Harris Children's Self-Concept Scale.* Nashville, Tennessee: Counselor Recordings and Tests.

Piers, E. V., and Harris, D. B. 1964. Age and other correlates of self-concept in children. *Journal of Educational Psychology* 55: 91–95.

Poe, E. A. 1938. *The Complete Tales and Poems of Edgar Allan Poe.* New York: Modern Library.

Powell, G. J., and Fuller, M. 1970. School desegregation and self-concept. Paper presented at 47th Annual Meeting of the American Orthopsychiatric Association in San Francisco.

———. 1973. *Black Monday's Children.* New York: Appleton-Century-Crofts.

Proshansky, H., and Newton, P. 1968. The nature and meaning of Negro self-identity. In M. Deutsch, I. Katz, and A. R. Jensen, eds. *Social Class, Race, and Psychological Development.* New York: Holt, Rinehart and Winston. Pp. 178–218.

Purkey, W. W. 1970. *Self-Concept and School Achievement.* Englewood Cliffs, New Jersey: Prentice-Hall.

Raimy, V. C. 1948. Self-reference in counseling interviews. *Journal of Consulting Psychology* 12: 153–163.

Rainwater, L. 1966. Crucible of identity: the Negro lower-class family. *Daedalus* 95: 172–216.

Reeder, L. G., Donohue, G. A., and Biblarz, A. 1960. Conceptions of self and others. *American Journal of Sociology* 66: 153–159.

Reisman, D., Glazer, N., and Denney, R. 1950. *The Lonely Crowd.* New Haven: Yale University Press.

Rogers, C. R. 1951. *Client-centered Therapy: Its Current Practice, Implications, and Theory.* Boston: Houghton-Mifflin.

Rose, A. M., ed. 1962. *Human Behavior and Social Processes.* Boston: Houghton-Mifflin.

Rosen, B. C. 1959. Race, ethnicity, and the achievement syndrome. *American Sociological Review* 24: 47–60.

Rosenberg, F., and Simmons, R. G. 1975. Sex differences in the self-concept in adolescence. *Sex Roles: A Journal of Research* 1: 147–159.

Rosenberg, M. 1957. *Occupations and Values*. Glencoe, Illinois: The Free Press.

———. 1962a. The dissonant religious context and emotional disturbance. *American Journal of Sociology* 68: 1–10.

———. 1962b. Test factor standardization as a method of interpretation. *Social Forces* 41: 53–61.

———. 1965. *Society and the Adolescent Self-Image*. Princeton: Princeton University Press.

———. 1968. *The Logic of Survey Analysis*. New York: Basic Books.

———. 1973a. Which significant others? *American Behavioral Scientist* 16: 829–860.

———. 1973b. The logical status of suppressor variables. *Public Opinion Quarterly* 37: 359–372.

Rosenberg, M., and Simmons, R. G. 1972. *Black and White Self-Esteem: The Urban School Child*. Rose Monograph Series. Washington, D.C.: American Sociological Association.

Schilder, P. 1968. The image and appearance of the human body. In C. Gordon and K. J. Gergen, eds. *The Self in Social Interaction*. New York: Wiley. Pp. 107–114.

Schutz, A. 1970. *On Phenomenology and Social Relations*. Edited by Helmut R. Wagner. Chicago: University of Chicago Press.

Schwartz, M., and Stryker, S. 1971. *Deviance, Selves and Others*. Rose Monograph Series. Washington, D.C.: American Sociological Association.

Schwartz, M., Fearn, G., and Stryker, S. 1966. A note on self-conception and the emotionally disturbed role. *Sociometry* 29: 300–305.

Seeman, M. 1966. Status and identity: the problem of inauthenticity. *Pacific Sociological Review* 9: 67–73.

Seidman, J. 1962. Telephone workers. In S. Nosow and W. Form, eds. *Man, Work and Society*. New York: Basic Books. Pp. 493–504.

Sewell, W. H., and Armer, J. M. 1966. Neighborhood context and college plans. *American Sociological Review* 31: 159–168.

Shantz, C. 1975. The development of social cognition. In E. M. Hetherington, ed. *Review of Child Development Theory and Research*, Vol. 5. Chicago: University of Chicago Press.

Sherif, M. 1962. The self and reference groups: meeting ground of individual and group approaches. *Annals of the New York Academy of Sciences* 96: 797–813.

———. 1968. Self-concept. *International Encyclopedia of the Social Sciences*, Vol. 14. New York: Macmillan.

Sherif, M., and Cantril, H. 1947. *The Psychology of Ego-Involvements*. New York: Wiley.

Sherwood, J. J. 1965. Self identity and referent others. *Sociometry* 28: 66–81.

———. 1967. Increased self-evaluation as a function of ambiguous evaluations by referent others. *Sociometry* 30: 404–409.

Shibutani, T. 1961. *Society and Personality*. Englewood Cliffs, New Jersey: Prentice-Hall.

Siegel, S. 1956. *Nonparametric Statistics for the Behavioral Sciences*. New York: McGraw-Hill.

Silber, E., and Tippett, J. S. 1965. Self-esteem: clinical assessment and measurement validation. *Psychological Reports* 16: 1017–1071.

Simmel, G. 1950. The stranger. In K. Wolff, ed. *The Sociology of Georg Simmel*. New York: The Free Press. Pp. 402–408.

Simmons, R. G., Blyth, D. A., Brown, L., and Bush, D. E. 1977. Role of school environment and puberty on self-image development. Paper presented at the American Sociological Association Meetings, Chicago, September.

Simmons, R. G., Brown, L., Bush, D. E., and Blyth, D. A. 1977. Self-esteem and achievement of black and white early adolescents. Unpublished paper.

Simmons, R. G., and Rosenberg, F. 1975. Sex, sex roles, and self-image. *Journal of Youth and Adolescence* 4: 229–258.

# References

Simmons, R. G., and Rosenberg, M. 1971. Functions of children's perceptions of the stratification system. *American Sociological Review* 36: 235–249.

Simpson, R. L., and Simpson, I. H. 1959. The psychiatric attendant: development of an occupational self-image in a low-status occupation. *American Sociological Review* 24: 389–392.

Smith, M. B. 1950. The phenomenological approach in personality theory: some critical remarks. *Journal of Abnormal and Social Psychology* 45: 516–522.

———. 1968. Competence and socialization. In John Clausen, ed. *Socialization and Society*. Boston: Little, Brown.

Snyder, M. L., Stephan, W. G., and Rosenfield, D. 1976. Egotism and attribution. *Journal of Personality and Social Psychology* 33: 435–441.

Snygg, D., and Combs, A. W. 1949. *Individual Behavior: A New Frame of Reference for Psychology*. New York: Harper.

Soares, A. T., and Soares, L. M. 1969. Self-perceptions of culturally disadvantaged children. *American Educational Research Journal* 6: 31–45.

———. 1972. The self-concept differential in disadvantaged and advantaged students. *Proceedings of the Annual Convention of the American Psychological Association* 7: 195–196.

Stephan, C., Kennedy, J. C., and Aronson, E. 1977. The effects of friendship and outcome on task attribution. *Sociometry* 40: 107–112.

Stern, A. J., and Searing, D. D. 1976. The stratification beliefs of English and American adolescents. *British Journal of Political Science* 6: 177–201.

St. John, N. H. 1971. The elementary classroom as a frog pond: self-concept, sense of control and social context. *Social Forces* 49: 581–595.

———. 1975. *School Desegregation: Outcomes for Children*. New York: Wiley.

Stonequist, E. 1937. *The Marginal Man*. New York: Scribner's.

Stouffer, S. A., Suchman, E. A., DeVinney, L. C., Star, S. A., and Williams, R. M. 1949. *Adjustment During Army Life*. Princeton, New Jersey: Princeton University Press.

Stouffer, S. A., Guttman, L., Suchman, E. A., Lazarsfeld, P. F., Star, S. A., and Clausen, J. A. 1950. *Measurement and Prediction*. Princeton, New Jersey: Princeton University Press.

Stouffer, S. A., Borgatta, E. F., Hays, D. G., and Henry, A. F. 1952. A technique for improving cumulative scales." *Public Opinion Quarterly* 16: 273–291.

Streufert, S., and Streufert, S. E. 1969. Effects of conceptual structure, failure, and success on attribution of causality and interpersonal attitudes. *Journal of Personality and Social Psychology* 11: 138–147.

Strong, S. M. 1942. Social types in a minority group: formulation of a method. *American Journal of Sociology* 48: 563–573.

Stryker, S. 1962. Conditions of accurate role-taking: a test of Mead's theory. In A. M. Rose, ed. *Human Behavior and Social Processes*. Boston: Houghton-Mifflin. Pp. 41–62.

Sullivan, H. S. 1947. *Conceptions of Modern Psychiatry: The First William Alanson White Memorial Lectures*. Washington, D.C.: The William Alanson White Psychiatric Foundation.

———. 1953. *The Interpersonal Theory of Psychiatry*. New York: Norton.

Sweet, J. R., and Thornburg, K. R. 1971. Preschoolers' self and social identity within the family structure. *Journal of Negro Education* 40: 22–27.

Symonds, P. M. 1951. *The Ego and the Self*. New York: Appleton-Century-Crofts.

Taeuber, K. E., and Taeuber, A. F. 1965. *Negroes in Cities*. Chicago: Aldine.

Thibaut, J. W., and Kelley, H. 1959. *The Social Psychology of Groups*. New York: Wiley.

Tippett, J., and Silber, E. 1965. Self-image stability: the problem of validation. *Psychological Reports* 17: 323–329.

Tiryakian, E. S. 1968. The existential self and the person. In C. Gordon and K. J. Gergen, eds. *The Self in Social Interaction*. New York: Wiley. Pp. 75–86.

Trimble, J. E. 1974. Say goodbye to the Hollywood Indian: results of a nationwide survey of the self-image of the American Indian. Paper presented at the Annual Meeting of the American Psychological Association, New Orleans, August.

Trowbridge, N. T. 1970. Self-concept and socio-economic class. *Psychology in the Schools* 7: 304–306.

Turner, R. 1976. The real self: from institution to impulse. *American Journal of Sociology* 81: 989–1016.

Turner, R. H., and Vanderlippe, R. 1958. Self-ideal congruence as an index of adjustment. *Journal of Abnormal and Social Psychology* 57: 202–206.

Twain, M. 1922. *Pudd'nhead Wilson and Those Extraordinary Twins*. New York: Harper.

U. S. Bureau of the Census. 1972. *Census of Housing: 1970 Block Statistics*. Financial Report HC(3)-68, Chicago, Illinois-Northwestern Indiana Urbanized Area. Washington, D.C.: Government Printing Office.

Veblen, T. 1934. *The Theory of the Leisure Class*. New York: Modern Library. (orig. 1899, Mentor Books.)

Videbeck, R. 1960. Self-conception and the reaction of others. *Sociometry* 23: 351–359.

Waterbor, R. 1972. Experiential bases of the sense of self. *Journal of Personality* 40: 162–179.

Weber, M. 1952. *Ancient Judaism*. Translated and edited by H. H. Gerth and D. Martindale. Glencoe, Illinois: The Free Press.

Webster, M., and Sobieszek, B. 1974. *Sources of Self-Evaluation*. New York: Wiley.

Weidman, J. C., Phelan, W. T., and Sullivan, M. A. 1972. The influence of educational attainment on self-evaluations of competence. *Sociology of Education* 45: 303–312.

Weiner, B. 1974. *Achievement Motivation and Attribution Theory*. Morristown, New Jersey: General Learning Press.

Weissman, D. 1965. *Dispositional Properties*. Carbondale, Illinois: Southern Illinois University Press.

Wells, L. E., and Marwell, G. 1976. *Self-Esteem: Its Conceptualization and Measurement*. Beverly Hills: Sage Publications.

Williams, R. M. 1951. *American Society: A Sociological Interpretation*. New York: A. A. Knopf.

Wortman, C. B., Costanzo, P. R., and Witt, T. R. 1973. Effect of anticipated performance on the attributions of causality to self and others. *Journal of Personality and Social Psychology* 27: 372–381.

Wylie, R. 1961. *The Self-Concept: A Critical Survey of Pertinent Research Literature*. Lincoln, Nebraska: University of Nebraska Press.

———. 1968. The present status of self theory. In E. Borgatta and W. Lambert, eds. *Handbook of Personality Theory and Research*. New York: Rand McNally. Pp. 728–787.

———. 1974. *The Self-Concept: Revised Edition. Volume 1. A Review of Methodological Considerations and Measuring Instruments*. Lincoln, Nebraska: University of Nebraska Press.

———. 1978. *The Self-Concept: Revised Edition. Volume 2. Theory and Research on Selected Topics*. Lincoln, Nebraska: University of Nebraska Press.

Yancey, W. L., Rigsby, L., and McCarthy, J. D. 1972. Social position and self-evaluation: the relative importance of race. *American Journal of Sociology* 78: 338–359.

Yinger, J. M. 1961. Social forces involved in group identification or withdrawal. *Daedalus* 48: 247–262.

Zeller, R. A., and Carmines, E. G. 1976. Factor scaling, external consistency and the measurement of theoretical constructs. *Political Methodology* 3: 215–252.

Zirkel, P. A., and Moses, E. G. 1971. Self-concept and ethnic group membership among public school students. *American Educational Research Journal* 8: 253–265.

# INDEX

# Index

Emotions: associated with self-attitudes, 26; development of concept of emotional control, 213–14; in ego-extensions, 35–37; see also Depressive affect; Feelings

Environmental context: changes in, and self-concept disturbance, 230–35, 237; see also Contextual dissonance

Epps, E. C., 130

Epstein, S., 61

Erdelyi, M., 276

Eriksen, B. A., 27

Eriksen, C. W., 27

Erikson, E. H., 5, 6, 8, 123, 224, 225, 236, 237

Ethnicity, see Contextual dissonance; Minority status

Exceptionality, 30

Existential self: self-concept distinguished from, 8

Experience contexts: effects of, on self-esteem, 105

Expertise: areas of, of significant others, as sources of differential significance, 93–94

Extant self-concept, 9–33; content of, 9–17; differential importance of components of, 18–19; dimensions of, 22–33; ego-extensions and, 34–38 (see also Ego-extensions); gap between desired and, 44–45; structure of, 17–22

Exterior (and interior) components of self-concept: 195–223; communication and, 218–20; conceptualization and, 220–22; development of, 196–208; introspection and, 216–18; from percept to concept of, 208–14

Family structure, 107–8, 173

Farley, R., 108

Feigenbaum, R., 269

Festinger, L., 274

Fitch, G., 60

Flapan, D., 215

Flavell, J., 15

Franks, D., 6, 31

French, J. R. P., 274

Freud, A., 224

Freud, S., 6, 7, 55

Frog-pond effect, 69

Fromm, E., 47, 189

Fuller, M., 103, 154

Future self (desired self), 196, 203, 205–6, 209

Generalized other, 65–67

Gergen, K., 5, 56, 286

Gerth, H. H., 128

Gill, B., 56

Gillman, G. B., 155

Global self-esteem: academic ability and, 21, 69–70; age and, 254; group rejection and, 179–80; self-concept disturbance and, 225, 227, 229, 230, 236, 239; significant others and, 85; see also Self-esteem

Goffman, E., 13, 45, 286

Goodman, M. E., 20

Gordon, C., 3, 11, 16, 27, 152, 153

Gough, H. G., 29

Gray, J. S., 162

Grinker, R. R., 225

Group introjection, 186–89

Group prestige, 161–62

Group pride, 186–89, 268

Group ranking, 151–52, 263

Group rejection (disidentification), 300; contextual dissonance and, 122–23; group pride and introjection and, 186–89; psychological centrality of group and effects of, 181–86; self-rejection and, 177–81

Guterman, S., 115

Guttentag, M., 188

Guzman, L. P., 155

Gynther, M. D., 153

Haigh, G. V., 41

Hall, C. S., 6

Hall, G. S., 224

Hammersmith, 61

Harding, J., 161, 275

Hardt, R. H., 154

Harris, L., 165

Hatt, P. K., 263

Havighurst, R., 285

Healey, G. W., 155

Heider, F., 71

Heilbrun, A. B., 29

Heiss, J., 55, 130, 275

Hendrickson, 162

Hermalin, A. I., 108

Herman, S. N., 124

Hessler, R. M., 99

Hilgard, E. R., 6, 27, 55, 260

Hishiki, P. C., 155

Hodge, R. W., 13

Hollingshead, A. B., 129

Hoppe, 276

Horney, K., 7, 21–22, 38–40, 44, 273

Horowitz, E. L., 162

Hughes, E. C., 84

Hunt, D. D., 154

# Index

# Index

This seminal work confronts one of the most fundamental issues in social psychology: how individuals conceive of themselves. Morris Rosenberg here traces the processes through which people arrive at particular conceptions of themselves from childhood through adulthood, the way these conceptions are influenced by and influence experience, and the means by which the individual defends his concept of himself from the constant onslaughts of the surrounding world.

Based on solid empirical material, much of it drawn from the author's own research among children and adolescents but including considerable data on adults as well, *Conceiving the Self* establishes for the first time a general theory of the self-concept and the social-psychological principles underlying its formation. Although the crucial importance of the self-concept in our understanding an individual's thought and behavior has long been recognized, most social psychologists have tended to limit themselves to just one aspect, self-esteem, ignoring such other crucial aspects of the self-concept as stability,